T0295061

Quantum Computation and Quantum Information Simulation using Python

A gentle introduction

Series Editor: **Barry Garraway** (School of Mathematical and Physical Sciences, University of Sussex, UK)

About the series

The IOP Series in Quantum Technology is dedicated to bringing together the most up to date texts and reference books from across the emerging field of quantum science and its technological applications. Prepared by leading experts, the series is intended for graduate students and researchers either already working in or intending to enter the field. The series seeks (but is not restricted to) publications in the following topics:

- Quantum biology
- Quantum communication
- Quantum computation
- Quantum control
- Quantum cryptography
- Quantum engineering
- Quantum machine learning and intelligence
- Quantum materials
- Quantum metrology
- Quantum optics
- Quantum sensing
- Quantum simulation
- Quantum software, algorithms and code
- Quantum thermodynamics
- Hybrid quantum systems

A full list of titles published in this series can be found here: https://iopscience.iop.org/bookListInfo/iop-series-in-quantum-technology.

Quantum Computation and Quantum Information Simulation using Python

A gentle introduction

Shinil Cho

Department of Physics, La Roche University, Pittsburgh, PA, USA

IOP Publishing, Bristol, UK

ISBN   978-0-7503-3963-6 (ebook)
ISBN   978-0-7503-3961-2 (print)
ISBN   978-0-7503-3964-3 (myPrint)
ISBN   978-0-7503-3962-9 (mobi)

DOI   10.1088/978-0-7503-3963-6

Version: 20220701

IOP ebooks

British Library Cataloguing-in-Publication Data: A catalogue record for this book is available from the British Library.

Published by IOP Publishing, wholly owned by The Institute of Physics, London

IOP Publishing, Temple Circus, Temple Way, Bristol, BS1 6HG, UK

US Office: IOP Publishing, Inc., 190 North Independence Mall West, Suite 601, Philadelphia, PA 19106, USA

To My family, Mari, Felix, and Michael.

Contents

Preface

In 2000, the National Science Foundation predicted that '*Quantum Information science (QIS) will have an extensive eventual impact on how science is taught at the college and secondary level, and will bring a deeper understanding of quantum physics to a broad segment of the lay public.*' Now, everyone talks about quantum computers as the future of computational means. Although there are many journal articles, books, and research articles on quantum computation and information, they are often either too easy or too difficult for physics students. There are many excellent lecture notes, textbooks, and videos available from highly recognized universities worldwide, but they are intensive and may not be ideal to those who want to self-study the subject and explore the current status of the quantum computation and information theory.

This book aims for an easy-to-read, self-paced but practical introduction to serious learners by taking a step-by-step approach without the need of rigorous mathematics. It does not describe what you can do with quantum computers but instead describes what other books do not write out, especially such content that may not be obvious for those who just started learning quantum computation. Fundamental concepts of physics behind quantum gates and selected programming examples of simulated quantum computation are explained in detail without referring to many other books and articles. This book intends to serve as a gateway to acquire more advanced knowledge and even using existing cloud-based quantum computers. Readers should take pencil and paper to follow the steps in order to feel the spin dynamics through the quantum gates.

Chapter 1 summarizes concepts and rules of vectors and matrices used in this book, and fundamental knowledge of quantum mechanics. They are applied throughout this book. Chapter 2 describes the classical binary gates and concepts of alternative universal gates that triggered quantum computation. Chapter 3 shows various quantum gates used to construct quantum circuits, and demonstrates how to implement the quantum gates and algorithms using Blueqat, a practical quantum gates simulation using Python. However, no prior knowledge of Python programming is required for investigating quantum circuits. Chapter 4 explains signature quantum algorithms so that the readers may acquire working knowledge of quantum circuits. Chapter 5 describes Bells' inequality for quantum entanglement and a scheme of quantum teleportation using the entanglement. Chapter 6 demonstrates how to send a code safely using photon quantum bits without being tampered with, called the BB84 protocol. The appendix briefly introduces the latest commercial advancements developed by Amazon, IBM, Microsoft, Google, and D-wave, and other companies.

You will be amazed to learn how much this cutting-edge technology has progressed once you immerse yourself into the subject. Let us dive into a practical gateway from this book.

Author biography

Shinil Cho

Shinil Cho attended Rikkyo University (St. Paul University) in Tokyo, Japan for his BS degree, Seoul National University in Seoul, Korea for MS, and the Ohio State University for PhD. He held post-doctoral fellowships at the Ohio State University and University of Florida, a visiting professor at University of South Carolina. He has been at La Roche University since 1995. Currently he is a full Professor at La Roche.

He has conducted research in critical phenomena of random systems, cryogenic magnetic resonance spectroscopy below 1 Kelvin, and biometric fingerprint authentication. His current research interest includes Monte Carlo analysis of random systems, quantum computation, biometrics, and physics education. Other than physics, he has many publications and presentations on biometrics in London, Gothenburg, Tokyo, Seoul, Hong Kong, Singapore, and several cities in the United States.

Acknowledgments

The author would like to thank Dr John Navas, the editor, at IOP Publishing for encouraging me with an opportunity to write this book, and Ms Phoebe Hooper and her staff for brushing up the text and careful proofreading. Without their help, this book would not be published.

IOP Publishing

Quantum Computation and Quantum Information Simulation using Python
A gentle introduction
Shinil Cho

Chapter 1

Two-level quantum systems

We need a basic knowledge of linear algebra (vectors and matrices) throughout this book. Because readers of this book may be unfamiliar with vectors and matrices, in the first half of this chapter we will briefly describe them. In the second half, rotations of spins and the coordinates, projection operators to represent observations, and entanglement and superposition of quantum states will be discussed. Mathematics of photon-based qbit is also given.

1.1 Vectors and matrices

1.1.1 Calculation rules of vectors and matrices

Two-dimensional vector

In this book, we use Dirac's vector notation used in quantum physics. Consider an arbitrary two-dimensional vector $|v>$ as shown in figure 1.1 in the *ket* vector format: $|v\rangle = \begin{bmatrix} v_1 \\ v_2 \end{bmatrix}$ where v_1 and v_2 are the x- and y-components, respectively. For a Euclidean space, the coefficients are real values.

Using a set of unit vectors, $|e_1 = \begin{bmatrix} 1 \\ 0 \end{bmatrix}$ and $|e_2\rangle = \begin{bmatrix} 0 \\ 1 \end{bmatrix}$, the vector $|v>$ can be expressed as

$$|v\rangle = v_1|e_1\rangle + v_2|e_2\rangle, \text{ or } |v\rangle = v_1\begin{bmatrix} 1 \\ 0 \end{bmatrix} + v_2\begin{bmatrix} 0 \\ 1 \end{bmatrix} = \begin{bmatrix} v_1 \\ v_2 \end{bmatrix}. \tag{1.1}$$

Define the *bra vector*: $\langle u|=[u_1 u_2]$, and the *inner (scalar) product* of two vectors, $<u|$ and $|v>$, can be defined as

$$\langle u|v\rangle = u_1 v_1 + u_2 v_2 \text{ (bracket!)}. \tag{1.2}$$

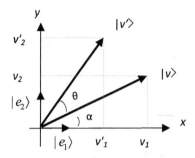

NOTE: $|v'\rangle$ is the rotated vector $|v\rangle$ by angle θ.

Figure 1.1. Two-dimensional vector space.

Using the inner product, the orthonormal property of the unit vectors can be expressed as

$$\langle e_i|e_j\rangle = \delta_{ij} \text{ where } \delta_{ij} = 1 \text{ if } i = j; \text{ and } 0 \text{ if } i \neq j \text{ (Kronecker's delta).} \quad (1.3)$$

Also, the definiton of the *outer product* of the two vectors is given by

$$|v\rangle\langle u| = \begin{bmatrix} v_1 \\ v_2 \end{bmatrix}[u_1 \quad u_2] = \begin{bmatrix} v_1u_1 & v_1u_2 \\ v_2u_1 & v_2u_2 \end{bmatrix}. \quad (1.4)$$

Matrices as operators on vectors

Because of the orthonormal property of the unit vectors, the vector components, v_1 and v_2, can be given by the inner products of the unit vector and the vector $|v\rangle$:

$$\langle e_1|v\rangle = v_1\langle e_1|e_1\rangle + v_2\langle e_1|e_2\rangle = v_1, \text{ and } \langle e_2|v\rangle = v_1\langle e_2|e_1\rangle + v_2\langle e_2|e_2\rangle = v_2. \quad (1.5)$$

Therefore,

$$|v\rangle = \langle e_1|v\rangle|e_1\rangle + \langle e_2|v\rangle|e_2\rangle = |e_1\rangle\langle e_1|v\rangle + |e_2\rangle\langle e_2|v\rangle. \quad (1.6)$$

Here, the outer products of the unit vectors, $|e_1\rangle\langle e_1|$ and $|e_2\rangle\langle e_2|$, are called the *projection operators*:

$$\hat{P}_1 = |e_1\rangle\langle e_1| = \begin{bmatrix} 1 & 0 \\ 0 & 0 \end{bmatrix}, \text{ and } \hat{P}_2 = |e_2\rangle\langle e_2| = \begin{bmatrix} 0 & 0 \\ 0 & 1 \end{bmatrix}. \quad (1.7)$$

Notice $\hat{P}_1|v\rangle = v_1|e_1\rangle$, $\hat{P}_2|v\rangle = v_2|e_2\rangle$, $\hat{P}_1|v\rangle + \hat{P}_2|v\rangle = |v\rangle$, and thus

$$\hat{P}_1 + \hat{P}_2 = \hat{I} \quad (1.8)$$

where $\hat{I}$ is the 2×2 unit matrix.

Below is a summary of the addition and multiplication of matrices. Suppose

$$A = \begin{bmatrix} a_{11} & a_{12} \\ a_{21} & a_{22} \end{bmatrix} \text{ and } B = \begin{bmatrix} b_{11} & b_{12} \\ b_{21} & b_{22} \end{bmatrix},$$

the basic matrix calculation rules are:

(1)

$$\delta A = \begin{bmatrix} \delta a_{11} & \delta a_{12} \\ \delta a_{21} & \delta a_{22} \end{bmatrix} \text{ where } \delta \text{ is a scalar constant,} \tag{1.9}$$

(2) Addition:

$$A \pm B = \begin{bmatrix} a_{11} \pm b_{11} & a_{12} \pm b_{12} \\ a_{21} \pm b_{21} & a_{22} \pm b_{22} \end{bmatrix}, \tag{1.10}$$

(3) Multiplication:

$$AB = \begin{bmatrix} a_{11}b_{11} + a_{12}b_{21} & a_{11}b_{12} + a_{12}b_{22} \\ a_{21}b_{11} + a_{22}b_{21} & a_{21}b_{12} + a_{22}b_{22} \end{bmatrix}. \tag{1.11}$$

Rotation and translation of a vector can be performed by applying corresponding matrices. For example, as shown in figure 1.1, rotating a vector $|v>$ by angle θ to create a new vector, $|v'>$, is expressed by $|v'> = R(\theta)|v>$. The rotational matrix, $R(\theta)$, can be obtained as follows. Let $v_1 = v \cos \alpha$ and $v_2 = v \sin \alpha$. The components of the rotated vector are given by

$$v_1' = v \cos(\alpha + \theta) = v \cos\alpha \cos\theta - v \sin\alpha \sin\theta = v_1 \cos\theta - v_2 \sin\theta,$$

and

$$v_2' = v \sin(\alpha + \theta) = v \sin\alpha \cos\theta + v \cos\alpha \sin\theta = v_2 \cos\theta + v_1 \sin\theta.$$

In the matrix representation (equation (1.11)), the above equations can be expressed as

$$\begin{bmatrix} v_1' \\ v_2' \end{bmatrix} = \begin{bmatrix} v_1 \cos\theta - v_2 \sin\theta \\ v_1 \sin\theta + v_2 \sin\theta \end{bmatrix} = \begin{bmatrix} \cos\theta & -\sin\theta \\ \sin\theta & \cos\theta \end{bmatrix} \begin{bmatrix} v_1 \\ v_2 \end{bmatrix},$$

and the rotational matrix is, thus, given by

$$R(\theta) = \begin{bmatrix} \cos\theta & -\sin\theta \\ \sin\theta & \cos\theta \end{bmatrix}. \tag{1.12}$$

Notice that rotation of the two-dimensional coordinate system (x, y) by angle θ to another coordinate system, (X, Y), is mathematically equivalent to the vector rotation by angle $-\theta$. Thus,

$$\begin{bmatrix} X \\ Y \end{bmatrix} = \begin{bmatrix} \cos\theta & \sin\theta \\ -\sin\theta & \cos\theta \end{bmatrix} \begin{bmatrix} x \\ y \end{bmatrix}. \tag{1.13}$$

1.1.2 Combining two different vector spaces—direct product

Consider two different two-dimensional vector spaces where each vector space has its own set of unit vectors or an orthonormal basis. Suppose a vector in each vector space is respectively given by

$$|u\rangle = \begin{bmatrix} u_1 \\ u_2 \end{bmatrix} \text{ and } |v\rangle = \begin{bmatrix} v_1 \\ v_2 \end{bmatrix}.$$

A vector in the coupled vector space is expressed by the *direct product* of the two vectors:

$$|u\rangle \otimes |v\rangle = \begin{bmatrix} u_1|v\rangle \\ u_2|v\rangle \end{bmatrix} = \begin{bmatrix} u_1v_1 \\ u_2v_2 \\ u_2v_1 \\ u_2v_2 \end{bmatrix}. \tag{1.14}$$

The coupled vector space becomes four-dimensional, and the coupled orthonormal basis is given by

$$|0\rangle \otimes |0\rangle = \begin{bmatrix} 1 \\ 0 \\ 0 \\ 0 \end{bmatrix}, \ |0\rangle \otimes |1\rangle = \begin{bmatrix} 0 \\ 1 \\ 0 \\ 0 \end{bmatrix}, \ |1\rangle \otimes |0\rangle = \begin{bmatrix} 0 \\ 0 \\ 1 \\ 0 \end{bmatrix}, \ |1\rangle \otimes |1\rangle = \begin{bmatrix} 0 \\ 0 \\ 0 \\ 1 \end{bmatrix}. \tag{1.15}$$

Notice that the direct product has the following distribution rule:

$$(|u_1\rangle + |u_2\rangle)_A \otimes (|v_1\rangle + |v_2\rangle)_B = |u_1\rangle_A|v_1\rangle_B + |u_1\rangle_A|v_2\rangle_B + |u_2\rangle_A|v_1\rangle_B + |u_2\rangle_A|v_2\rangle_B, \tag{1.16}$$

where the subscripts, A and B, indicate two different vector spaces. Notice that we often write $|u\rangle \otimes |v\rangle = |u\rangle|v\rangle$ for short. It is important not to change the order of vector for the direct product.

We also define the direct product of two matrices, $A = \begin{bmatrix} a_{11} & a_{12} \\ a_{21} & a_{22} \end{bmatrix}$ and $B = \begin{bmatrix} b_{11} & b_{12} \\ b_{21} & b_{22} \end{bmatrix}$, as

$$A \otimes B = \begin{bmatrix} a_{11}B & a_{12}B \\ a_{21}B & a_{22}B \end{bmatrix} = \begin{bmatrix} a_{11}b_{11} & a_{11}b_{12} & a_{12}b_{11} & a_{12}b_{12} \\ a_{11}b_{21} & a_{11}b_{22} & a_{12}b_{21} & a_{12}b_{22} \\ a_{21}b_{11} & a_{21}b_{12} & a_{22}b_{11} & a_{22}b_{12} \\ a_{21}b_{21} & a_{21}b_{22} & a_{22}b_{21} & a_{22}b_{22} \end{bmatrix}. \tag{1.17}$$

We will use the direct products when we discuss universal operators in chapter 2.

1.2 Foundation of quantum mechanics

1.2.1 General properties of quantum states

Here is a summary of properties that quantum systems must satisfy [1]. Whenever we create a quantum algorithm, it must satisfy these conditions.

(1) The quantum states of a system can be described by a single-valued continuous complex-valued function, $|\psi\rangle$, and the physics observables, A,

including energy, position, momentum, angular momentum, spins, and particle creation/annihilation, can be expressed as a mathematical operator, $\hat{A}$. Measurement values, $\{\varepsilon_k; k=1, 2, ...\}$, of the observable A satisfy the equation, $\hat{A}|\psi_k\rangle = \varepsilon_k|\psi_k\rangle$, where ε_k is the eigen value, and $|\psi_k\rangle$ is eigen function of the equation. When the observable is Hamiltonian (energy), the equation is called the Schrödinger equation, $\hat{H}|\psi_k\rangle = \varepsilon_k|\psi_k\rangle$.

(2) A single measurement of the physical observable A yields one of the possible eigen values $\{\varepsilon_k\}$. The equation itself does not determine the eigen functions and the eigen values without a boundary condition.

(3) A quantum state can be normalized, and its operators are linear:

  (i) $\langle\psi|\psi\rangle = 1$ where $\langle\psi|=(|\psi\rangle)^\dagger$ is the conjugate transpose of $|\psi>$, and

  (ii) $\hat{A}(a|\psi_1\rangle + b|\psi_2\rangle) = a\hat{A}|\psi_1\rangle + b\hat{A}|\psi_2\rangle$ where the coefficients a and b are constants of complex values.

(4) Because the physical observables are real values, the eigen values must be real numbers. Therefore, the operator $\hat{A}$ must be Hermitian, i.e., $\hat{A}^\dagger = \hat{A}$. Proof: Suppose $\hat{A}|\psi_k\rangle = \varepsilon_k|\psi_k\rangle$, and the eigen value is real, i.e., $\varepsilon_k = \varepsilon_k^*$. $\langle\psi_k|\hat{A}|\psi_k\rangle = \langle\psi_k|\varepsilon_k|\psi_k\rangle = \varepsilon_k\langle\psi_k|\psi_k\rangle = \varepsilon_k$, and $(\langle\psi_k|\hat{A}|\psi_k\rangle)^\dagger = \langle\psi_k|\hat{A}^\dagger|\psi_k\rangle = \varepsilon_k^*$. Thus, if the eigenvalue is a real number, $\hat{A}^\dagger = \hat{A}$.

(5) The eigen functions are orthogonal, i.e., $\langle\psi_i|\psi_j\rangle = 0$ if $i \neq j$. Proof. Suppose $\hat{A}|\psi_i\rangle = \varepsilon_i|\psi_i\rangle$ and $\hat{A}|\psi_j\rangle = \varepsilon_j|\psi_j\rangle$ where $\hat{A}$ is Hermitian. Notice $(\hat{A}|\psi_i\rangle)^\dagger = \langle\hat{A}^\dagger\psi_i| = \langle\hat{A}\psi_i|$ because $\hat{A}$ is Hermitian. Also notice that because the eigen values are real, $(\hat{A}|\psi_i\rangle)^\dagger = (\varepsilon_i|\psi_i\rangle)^\dagger = \varepsilon_i\langle\psi_i|$. Therefore, $\langle\hat{A}^\dagger\psi_i|\psi_j\rangle = \langle\psi_i|\hat{A}\psi_j\rangle = \varepsilon_j\langle\psi_i|\psi_j\rangle$, and $\langle\hat{A}^\dagger\psi_i|\psi_j\rangle - \langle\psi_i|\hat{A}\psi_j\rangle = (\varepsilon_i - \varepsilon_j)\langle\psi_i|\psi_j\rangle = 0$. Thus, $\langle\psi_i|\psi_j\rangle = 0$ unless $\varepsilon_i = \varepsilon_j$.

(6) The complete set of eigen functions $\{|\psi_k\rangle, k=1, 2, 3, ...\}$ forms an orthonormal basis. In other words, any quantum state of the given system can be expressed by the superposition of eigen functions: $|\psi\rangle = \sum_k c_k|\psi_k\rangle$

where $\langle\psi_j|\psi_k\rangle = \delta_{jk}$ and $c_k = \langle\psi_k|\psi\rangle$.

(7) Suppose a coordinate system for observation is changed to another coordinated system, a complete set of the eigen functions $\{|\psi_k\rangle, k=1, 2, 3, ...\}$ is transformed to another set of eigen functions $\{|\varphi_k\rangle, k=1, 2, 3, ...\}$, i.e., $|\varphi\rangle = \hat{U}|\psi\rangle$ where $\hat{U}$ is the operator for this transformation. Then, $|\varphi\rangle_j = \sum_{k=1} u_{jk}|\psi\rangle_k$, and $\langle\varphi_i|\varphi_j\rangle = \sum_{k,m}(u_{ik})^\dagger u_{mj}\langle\psi_k|\psi_m\rangle$. Since both sets of eigen functions are orthonormal, $\langle\varphi_i|\varphi_j\rangle = \delta_{i,j} = \sum_{k,m}(u_{ik})^\dagger u_{mj}\delta_{k,m}$.

Note: An operator, which satisfies $\sum_k(u_{ik})^\dagger u_{kj} = \delta_{i,j}$, is called the unitary operator, and $\hat{U}^\dagger\hat{U} = \hat{I}$, i.e., $\hat{U}^\dagger = \hat{U}^{-1}$. Operators that change quantum states must be unitary.

(8) The expectation value of the observable A is given by $\langle A\rangle = \langle\psi|\hat{A}|\psi\rangle$. In particular, $\langle A\rangle = \langle\psi_k|\hat{A}|\psi_k\rangle = \varepsilon_k$ if $|\psi\rangle = |\psi_k\rangle$.

(9) A projection operator is given by $\hat{P}_k = |\psi_k\rangle\langle\psi_k|$. If $|\psi\rangle = \sum_k c_k|\psi_k\rangle$, then $\hat{P}_k|\psi\rangle = \sum_j c_j|\psi_k\rangle\langle\psi_k|\psi_j\rangle = \sum_j c_j|\psi_k\rangle\delta_{kj} = c_k|\psi_k\rangle$.

The probability of observing the eigen state $|\psi_k\rangle$ is

$$|\langle\psi|\hat{P}_k|\psi\rangle|^2 = |\langle\psi|\sum_j c_j|\psi_k\rangle\langle\psi_k|\psi_j\rangle|^2 = |\langle\psi|\sum_j c_j|\psi_k\rangle\delta_{kj}|^2 = |c_k\langle\psi|\psi_k\rangle|^2 = |c_k|^2.$$

(10) Matrix representation of operator $\hat{U}$.

Let $|\psi\rangle = \sum_j c_j|\psi_k\rangle$ where $c_j = \langle\psi_j|\psi\rangle$, and $|\psi'\rangle = \hat{U}|\psi\rangle = \sum_j c_j\hat{U}|\psi_j\rangle$.

If $|\psi'\rangle = \sum_j c'_j|\psi_k\rangle$ where $c'_j = \langle\psi_j|\psi'\rangle$, then $|\psi'\rangle$ becomes

$|\psi'\rangle = \sum_i c'_i|\psi_i\rangle = \sum_i c_i\hat{U}|\psi_i\rangle$, and thus,

$$c'_i = \langle\psi_i|\psi'\rangle = \langle\psi_i|\hat{U}\psi\rangle = \langle\psi_i|\sum_j c_j\hat{U}\psi_j\rangle = \sum_j \langle\psi_i|\hat{U}|\psi_j\rangle c_j.$$

Define $\langle\psi_i|\hat{U}|\psi_j\rangle = U_{ij}$, and we obtain $c'_i = \sum_j U_{ij}c_j$. Therefore, the transform

$|\psi\rangle \rightarrow |\psi'\rangle = \hat{U}|\psi\rangle$ can be represented by a linear transform of $\{c_i\}$: $c'_i = \sum_j U_{ij}c_j$.

1.3 Quantum state vectors

Without rigorous mathematics, we describe quantum states in an analogy of the two-dimensional vectors described above. Here, we focus on two-level systems to describe a quantum bit (qbit). Readers may wonder whether the term, quantum bit, should be written as 'qubit' or 'qbit.' Refer to an interesting article on this subject.[1]

1.3.1 Two-level quantum state vector: qbit

An electron is known to have two possible spin states. It can be described on a two-dimensional complex vector space. When we measure the spin state along the z-axis or the 'vertical' axis, which is the default direction of external magnetic field, we observe the spin up (↑) or down (↓) state. We assign the following vectors |0> and |1> to the spin states:

$$\text{Spin up: } |\uparrow\rangle = |0\rangle = \begin{bmatrix} 1 \\ 0 \end{bmatrix}, \text{ and spin down: } |\downarrow\rangle = |1\rangle = \begin{bmatrix} 0 \\ 1 \end{bmatrix}. \qquad (1.18)$$

The spin state vectors form an orthonormal basis of the vector space, $\{|0\rangle, |1\rangle\}$ where $\langle 0|0\rangle = \langle 1|1\rangle = 1$, and $\langle 0|1\rangle = \langle 1|0\rangle = 0$. An arbitrary spin state can be given by $|\psi\rangle = a|0\rangle + b|1\rangle$ where the coefficients a and b are complex value constants, and the spin state is normalized, $|\langle\psi|\psi\rangle| = |a|^2 + |b|^2 = 1$ because either spin state will be observed by a measurement. The spin state $|\psi\rangle$ is called a qbit in the quantum computation/information. The complex coefficients, a and b, can be expressed as

---

[1] There is an interesting explanation for qbit and qubit. See https://scienceblogs.com/pontiff/2007/11/27/qubit-qbit-qbit-or-qbert-1)

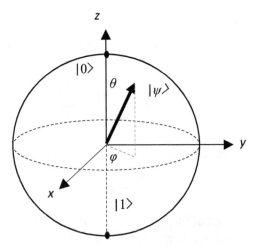

Figure 1.2. Bloch sphere.

$a = \cos\left(\dfrac{\theta}{2}\right)$ and $b = e^{i\varphi} \sin\left(\dfrac{\theta}{2}\right)$, using phase angles, φ and θ, which satisfy the normalization condition, $|a|^2 + |b|^2 = 1$. A Bloch sphere, shown in figure 1.2, is often used to visualize the qbit dynamics although we will not use it in this book.

1.3.2 Projection operators for spin states

If one observes the spin state, $|\psi\rangle = a|0\rangle + b|1\rangle$, along the z-axis, the measurement yields either spin up or down. Observation of the spin state can be interpreted as the application of the projection operators. From equation (1.7), the projection operators for the spin state described by the orthonormal basis, $\{|0\rangle, |1\rangle\}$, are given

$$\hat{P}_0 = |0\rangle\langle 0| = \begin{bmatrix} 1 \\ 0 \end{bmatrix}[1 \ \ 0] = \begin{bmatrix} 1 & 0 \\ 0 & 0 \end{bmatrix} \text{ and } \hat{P}_1 = |1\rangle\langle 1| = \begin{bmatrix} 0 \\ 1 \end{bmatrix}[0 \ \ 1] = \begin{bmatrix} 0 & 0 \\ 0 & 1 \end{bmatrix}. \quad (1.19)$$

The probability of observing the spin up state along the z-axis, $|a|^2$, can be calculated from a measurement

$$\hat{P}_0|\psi\rangle = |0\rangle\langle 0|\psi\rangle = |0\rangle\langle 0|(a|0\rangle + b|1\rangle) = a\begin{bmatrix} 1 & 0 \\ 0 & 0 \end{bmatrix}\begin{bmatrix} 1 \\ 0 \end{bmatrix} + b\begin{bmatrix} 1 & 0 \\ 0 & 1 \end{bmatrix}\begin{bmatrix} 0 \\ 1 \end{bmatrix} = a\begin{bmatrix} 1 \\ 0 \end{bmatrix} = a|0\rangle, \quad (1.20)$$

and the probability of observing the spin down state along the z-axis, $|b|^2$, can be calculated from a measurement

$$\hat{P}_1|\psi\rangle = |1\rangle\langle 1|\psi\rangle = |1\rangle\langle 1|(a|0\rangle + b|1\rangle) = a\begin{bmatrix} 0 & 0 \\ 0 & 1 \end{bmatrix}\begin{bmatrix} 1 \\ 0 \end{bmatrix} + b\begin{bmatrix} 0 & 0 \\ 0 & 1 \end{bmatrix}\begin{bmatrix} 0 \\ 1 \end{bmatrix} = b\begin{bmatrix} 0 \\ 1 \end{bmatrix} = b|1\rangle. \quad (1.21)$$

Once a measurement is conducted on a quantum system, the quantum state of the system is collapsed, and a successive measurement yields the same state as the previous measurement. This means

$$\hat{P}_0\hat{P}_0 = \hat{P}_0, \ \ \hat{P}_1\hat{P}_1 = \hat{P}_1, \text{ and } \hat{P}_0\hat{P}_1 = \hat{P}_1\hat{P}_0 = 0.$$

1.3.3 Time evolution of spin states

To obtain the time evolution and rotation of spins, we use exponential operators. We define the form of exponential operators as

$$e^{u\hat{A}} = \exp(u\hat{A}) \equiv \sum_{n=0}^{\infty} \frac{1}{n!}(u\hat{A})^n = \sum_{n=0}^{\infty} \frac{u^n}{n!}\hat{A}^n \tag{1.22}$$

where u is a complex number. Suppose the wave function is given by $|\psi(0)\rangle$ at $t = 0$, the wave function satisfies the time dependent Schrödinger equation:

$$i\hbar\frac{\partial|\psi(t)\rangle}{\partial t} = \hat{H}|\psi(t)\rangle$$

where $\hat{H}$ is the Hamiltonian of the system. From the Schrödinger equation, the time evolution of the wave function is given by

$$|\psi(t)\rangle = \exp\left(-\frac{i}{\hbar}\hat{H}t\right)|\psi(0)\rangle. \tag{1.23}$$

Proof.

$$i\hbar\frac{\partial|\psi(t)\rangle}{\partial t} = \frac{\partial}{\partial t}\left(e^{-\frac{i}{\hbar}\hat{H}t}|\psi(0)\rangle\right) = i\hbar\frac{\partial}{\partial t}\left[\sum_{n=0}^{\infty}\frac{1}{n!}\left(-\frac{i\hat{H}t}{\hbar}\right)^n\right]|\psi(0)\rangle$$

$$= \hat{H}\sum_{n=1}^{\infty}\frac{1}{(n-1)!}\left(-\frac{i\hat{H}t}{\hbar}\right)^{n-1}(n-1)t^{n-1}|\psi(0)\rangle$$

$$= i\hbar\sum_{n=1}^{\infty}\frac{1}{n!}\left(-\frac{i\hat{H}t}{\hbar}\right)^n(nt^{n-1})|\psi(0)\rangle$$

$$= -i\hbar\left(\frac{i\hat{H}}{\hbar}\right)\sum_{n=1}^{\infty}\frac{1}{(n-1)!}\left(-\frac{i\hat{H}t}{\hbar}\right)^{n-1}(n-1)t^{n-1}|\psi(0)\rangle$$

$$= \hat{H}e^{-\frac{i}{\hbar}\hat{H}t}|\psi(0)\rangle = \hat{H}|\psi(t)\rangle.$$

1.3.4 Rotation of spin states

Define the rotational operator, $\hat{R}_j(\delta\theta)$ of an infinitesimal angle $\delta\vec{\theta}$ about the rotational axis j, such that $\hat{R}_j(\delta\vec{\theta})|\psi(\vec{r})\rangle = |\psi'(\vec{r})\rangle$ on the wave function $|\psi(\vec{r})\rangle$. Because the position of the vector after it is rotated forward by an infinitesimal angle $\delta\vec{\theta}$ is given by $\vec{r}' = \vec{r} + \delta\vec{\theta} \times \vec{r}$ using the vector product (figure 1.3), we obtain

$$|\psi'(\vec{r})\rangle = |\psi(\vec{r} - \delta\vec{\theta} \times \vec{r})\rangle,$$

and thus,

$$\hat{R}_j(\delta\vec{\theta})|\psi(\vec{r})\rangle = |\psi(\vec{r} - \delta\vec{\theta} \times \vec{r})\rangle \approx |\psi(\vec{r})\rangle - (\delta\vec{\theta} \times \vec{r}) \cdot \nabla|\psi(\vec{r})\rangle$$

$$= |\psi(\vec{r})\rangle - \frac{i}{\hbar}(\delta\vec{\theta} \times \vec{r}) \cdot \vec{p}\,|\psi(\vec{r})\rangle = \left(1 - \frac{i}{\hbar}\delta\vec{\theta} \cdot \vec{L}\right)|\psi(\vec{r})\rangle \tag{1.24}$$

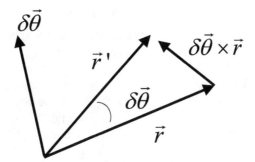

Figure 1.3. Infinitesimal rotation of the position vector.

where $\vec{p} = -i\hbar\nabla$ is the linear momentum, and $\vec{L} = \vec{r} \times \vec{p}$ is the angular momentum.

The infinitesimal angle is defined as $\delta\vec{\theta} = \lim\limits_{N\to\infty} \vec{\theta}/N$, and we obtain

$$\hat{R}_j(\vec{\theta}) = \lim_{N\to\infty}\left(1 - \frac{i}{\hbar}\frac{\vec{\theta}\cdot\vec{L}}{N}\right)^N = \exp(-i\vec{\theta}\cdot\vec{L}/\hbar). \tag{1.25}$$

The spin rotational operators about the x, the y, and the z axes are, respectively, given by replacing the angular momentum of each axis with Pauli's spin matrix of the corresponding axis:

$$\hat{R}_j(\vec{\theta}) = \exp\left(-i\frac{\theta_j \hat{L}_j}{\hbar}\right) = \exp\left(-i\frac{\theta_j \hat{\sigma}_j}{2}\right), \tag{1.26}$$

where $j = x$, y, or z, and

$$\hat{\sigma}_x = \begin{bmatrix} 0 & 1 \\ 1 & 0 \end{bmatrix}, \hat{\sigma}_y = \begin{bmatrix} 0 & -i \\ i & 0 \end{bmatrix}, \hat{\sigma}_z = \begin{bmatrix} 1 & 0 \\ 0 & -1 \end{bmatrix}, \text{ and } \hat{I} = \begin{bmatrix} 1 & 0 \\ 0 & 1 \end{bmatrix}.$$

Thus,

$$\hat{R}_x(\theta) = \exp\left(-i\frac{\theta\hat{\sigma}_x}{2}\right) = \left(\cos\frac{\theta}{2}\right)\hat{I} + i\left(\sin\frac{\theta}{2}\right)\hat{\sigma}_x = \begin{bmatrix} \cos(\theta/2) & -i\sin(\theta/2) \\ -i\sin(\theta/2) & \cos(\theta/2) \end{bmatrix}, \tag{1.27}$$

$$\hat{R}_y(\theta) = \exp\left(-i\frac{\theta\hat{\sigma}_y}{2}\right) = \left(\cos\frac{\theta}{2}\right)\hat{I} + i\left(\sin\frac{\theta}{2}\right)\hat{\sigma}_y = \begin{bmatrix} \cos(\theta/2) & -\sin(\theta/2) \\ \sin(\theta/2) & \cos(\theta/2) \end{bmatrix}, \tag{1.28}$$

$$\hat{R}_z(\theta) = \exp\left(-i\frac{\theta\hat{\sigma}_z}{2}\right) = \left(\cos\frac{\theta}{2}\right)\hat{I} + i\left(\sin\frac{\theta}{2}\right)\hat{\sigma}_z = \begin{bmatrix} e^{-i\theta/2} & 0 \\ 0 & e^{i\theta/2} \end{bmatrix}. \tag{1.29}$$

where we have omitted the suffix of the angles. For basic concepts of spin states, refer to a textbook on quantum mechanics [1].

1.3.5 Rotation of a spin observation coordinate frame

In order to observe a spin state, an external magnetic field is applied to the spin. If the external magnet is along the z-axis, we observe either |0> or |1> as we discussed above (equations (1.19) and (1.20)). Now, if we rotate the external magnetic field by angle θ about one of the coordinate axes, we rotate the observation coordinate frame. Thus, if we rotate the z-axis by angle θ on the zx-plane, i.e., if we rotate the 'vertical' spin-measurement coordinate frame about the y-axis by angle θ, we obtain a new 'tilted' measurement coordinate frame where the rotated orthonormal basis is given by:

$$|0'\rangle = \hat{R}_y(-\theta)|0\rangle = \begin{bmatrix} \cos(\theta/2) & \sin(\theta/2) \\ -\sin(\theta/2) & \cos(\theta/2) \end{bmatrix}\begin{bmatrix} 1 \\ 0 \end{bmatrix} = \begin{bmatrix} \cos(\theta/2) \\ -\sin(\theta/2) \end{bmatrix}, \qquad (1.30)$$

and

$$|1'\rangle = \hat{R}_y(-\theta)|1\rangle = \begin{bmatrix} \cos(\theta/2) & \sin(\theta/2) \\ -\sin(\theta/2) & \cos(\theta/2) \end{bmatrix}\begin{bmatrix} 0 \\ 1 \end{bmatrix} = \begin{bmatrix} \sin(\theta/2) \\ \cos(\theta/2) \end{bmatrix}.$$

In particular, if $\theta = \pi/2$, the direction of the external magnetic field becomes 'horizontal,' or

$$|0'\rangle = |\rightarrow\rangle = \begin{bmatrix} \cos(\pi/4) \\ -\sin(\pi/4) \end{bmatrix} = \begin{bmatrix} 1/\sqrt{2} \\ -1/\sqrt{2} \end{bmatrix}, \text{ and } |1'\rangle = |\leftarrow\rangle$$

$$= \begin{bmatrix} \sin(\pi/4) \\ \cos(\pi/4) \end{bmatrix} = \begin{bmatrix} 1/\sqrt{2} \\ 1/\sqrt{2} \end{bmatrix}. \qquad (1.31)$$

Recall that if we observe the spin state, $|\psi\rangle = a|0\rangle + b|1\rangle$, along the 'vertical' coordinates frame where the external magnetic field is along the z-axis, an observed state is either |0> or |1> with the corresponding probability $|a|^2$ or $|b|^2$. If we observe the spin state using the 'horizontal' measurement frame where the external magnetic field is along the x-axis, the spin state to be observed can be determined by applying the projection operators, $\hat{P}_{|0'\rangle} = |0'\rangle\langle 0'|$ and $\hat{P}_{|1'\rangle} = |1'\rangle\langle 1'|$ to the basis |0> and |1>:

$$\hat{P}_{|0'\rangle}|0\rangle = a(|0'\rangle\langle 0'|)|0\rangle = a\begin{bmatrix} 1/\sqrt{2} & 1/\sqrt{2} \end{bmatrix}\begin{bmatrix} 1 \\ 0 \end{bmatrix}|0'\rangle = \frac{a}{\sqrt{2}}|0'\rangle, \qquad (1.32)$$

and

$$\hat{P}_{|1'\rangle}|1\rangle = b(|1'\rangle\langle 1'|)|1\rangle = b\begin{bmatrix} 1/\sqrt{2} & -1/\sqrt{2} \end{bmatrix}\begin{bmatrix} 1 \\ 0 \end{bmatrix}|1'\rangle = \frac{b}{\sqrt{2}}|1'\rangle.$$

Thus, the spin state becomes $|\psi\rangle = \frac{1}{\sqrt{2}}(a|0'\rangle + b|1'\rangle)$ in the 'horizontal' measurement frame, and we observe |0'> or |1'>, and the probability of observing |0'> is $|a|^2/2$ and the probability of observing |1'> spin is $|b|^2/2$ along the horizontal magnetic field. This is a useful description when we discuss Bell's inequality where three different observation frames are involved (section 5.1).

1.4 Non-cloning principle for qbit

One of the distinct characteristics of a qbit is that it cannot be copied because an act of observation to copy a qbit, $|\psi\rangle = a|0\rangle + b|1\rangle$, changes the quantum state to either $|0\rangle$ or $|1\rangle$, and any successive measurement yields the same state as the first measurement. Let us restate this non-cloning principle.

Suppose we could define a 'copy' operator, $\hat{C}$, such that $\hat{C}|\psi\rangle = |\psi\rangle|\psi\rangle$ where $|\psi\rangle = a|0\rangle + b|1\rangle$. Then, by the definition of the copy operator,

$$\hat{C}|\psi\rangle = |\psi\rangle|\psi\rangle = (a|0\rangle + b|1\rangle)(a|0\rangle + b|1\rangle)$$
$$= a^2|0\rangle|0\rangle + ab(|0\rangle|1\rangle + |1\rangle|0\rangle) + b^2|1\rangle|1\rangle.$$

On the other hand,

$$\hat{C}|\psi\rangle = \hat{C}(a|0\rangle + b|1\rangle) = a\hat{C}|0\rangle + b\hat{C}|1\rangle = a|0\rangle|0\rangle + b|1\rangle|1\rangle. \tag{1.33}$$

Therefore, unless $a=0$ or $b=0$, we cannot define the copy operator. If $a=0$, then $b^2 = b$ and thus $b=1$; and if $b=0$, $a^2=a$ and $a=1$. These conditions indicate the pure quantum state $|\psi\rangle = |0\rangle$ or $|\psi\rangle = |1\rangle$ in which there is no quantum fluctuation. The non-cloning principle plays an important role in detecting data tapping while transmitting a secret code using the quantum teleportation (section 6.3).

1.5 Quantum entanglement

1.5.1 What is entanglement?

Suppose a set of two electron spins, spin A and spin B, are coupled, the net spin S is 1 or 0. Such an example is two electrons in a hydrogen molecule. As shown in table 1.1, if $S = 1$, then $S_z = +1$, 0, or -1 whereas, if $S=0$, then $S_z = 0$ only.

The quantum states of $S=1$ and $S_z = 0$ or $S= 0$ and $S_z = 0$ are called the *entangled* states. Suppose we conduct an observation of spin A when $S = 0$. Using the projection operator, the spin state becomes

$$\hat{P}_A|\psi\rangle = |0_A\rangle\langle 0_A|\left(\frac{1}{\sqrt{2}}(|0_A 1_B\rangle - |1_A 0_B\rangle)\right) = \frac{1}{\sqrt{2}}|0_A 1_B\rangle. \tag{1.34}$$

Table 1.1. Two coupled spins.

Net spin S	S_z	Quantum state
$S = 1$	+1	$\|0_A 0_B\rangle$
	0	$\frac{1}{\sqrt{2}}(\|0_A 1_B\rangle + \|0_A 1_B\rangle)$
	-1	$\|1_A 1_B\rangle$
$S = 0$	0	$\frac{1}{\sqrt{2}}(\|0_A 1_B\rangle - \|1_A 0_B\rangle)$

That is, if spin A is up ($\uparrow$), then spin B must be down ($\downarrow$). Observing the spin A will automatically determine the spin state B. It seems to be trivial, and this is also true for the classical physics. However, recall that unless we observe one of them, both spins A and B have equal probabilities of spin up *and* down!

What is important to perceive is that, in classical physics, all spin states are pre-determined regardless of whether we observe them or not. Therefore, the spin states A and B are already determined with the condition of the total spin value. On the other hand, in quantum physics, unless we make observation of spin A, spin A can be both up and down with equal probabilities, and spin B can also be both up and down with equal probabilities. This argument is very much like the quantum logic of Schrödinger's cat.

According to quantum mechanics, an act of observation of the spin A state instantly determines the spin B state, depending on the outcome of the spin state A. Remember that the spin A state is undermined before observation. This is why we call it the 'spooky' behavior of quantum entanglement. Furthermore, the entanglement is non-localizing, i.e., even if the spins are physically separated (without observation) by an astronomical distance after making the coupled state, they are still entangled!

The quantum entanglement is such a strange and unique quantum behavior, and it is one of the key properties that make quantum computation and information theory distinct from the classical theory. One of its astonishing examples is quantum teleportation, which will be discussed in chapter 5.

1.5.2 Superposition and entanglement

Imagine that two classical sinusoidal waves of frequencies, $\sin(\omega_1 t)$ and $\sin(\omega_2 t)$ are superposed, the resultant wave is given by

$$\sin(\omega_1 t) + \sin(\omega_2 t) = 2\cos\left(\frac{\omega_1 - \omega_2}{2}t\right)\sin\left(\frac{\omega_1 + \omega_2}{2}t\right).$$

There are two possible frequencies after superposition of two classical waves. The quantum superposition is fundamentally different from the superposition of n-classical waves which linearly gives only n different states. The number of states exponentially increases in the quantum superposition.

A qbit $|\psi\rangle = a|0\rangle + b|1\rangle$ is a superposition of |0> and |1> states. The possible states from superposition of two qbits are given by the direct product defined by equation (1.14)

$$|\psi\rangle = (a|0\rangle + b|1\rangle) \otimes (c|0\rangle + d|1\rangle) = ac|00\rangle + ad|01\rangle + bc|10\rangle + bd|11\rangle = \begin{bmatrix} ac \\ ad \\ bc \\ bd \end{bmatrix}. \quad (1.35)$$

Now, the number of possible states become four. If n-qbits are superposed,

$$|\psi\rangle \otimes |\psi\rangle \otimes \cdots \otimes |\psi\rangle$$

$$= \left(\frac{1}{\sqrt{2}}|0\rangle + \frac{1}{\sqrt{2}}|1\rangle\right) \otimes \left(\frac{1}{\sqrt{2}}|0\rangle + \frac{1}{\sqrt{2}}|1\rangle\right) \otimes \cdots \otimes \left(\frac{1}{\sqrt{2}}|0\rangle + \frac{1}{\sqrt{2}}|1\rangle\right)$$

$$= \frac{1}{\sqrt{2^n}}(|0\rangle \otimes |0\rangle \otimes \cdots \otimes |0\rangle + \cdots\cdots + |1\rangle \otimes |1\rangle \otimes \cdots \otimes |1\rangle) \quad (1.36)$$

$$= \frac{1}{\sqrt{2^n}}(|0\rangle|0\rangle\cdots|0\rangle + \cdots\cdots + |1\rangle|1\rangle\cdots|1\rangle) = \frac{1}{\sqrt{2^n}}(|00\rangle\cdots0\rangle + \cdots\cdots + |11\cdots1\rangle).$$

This represents a superposed 2^n-state from $|00\cdots0\rangle$ to $|11\cdots1\rangle$ with equal weight. Computation using superposed qbits is superior to the classical binary bit calculation (*quantum supremacy*) because we can perform computation on all the superposed states at once.

It must be noted that the quantum entanglement is not superposition of quantum states. If two qbits are entangled, they cannot be expressed as the direct product of two qbits. In other words, they cannot be separated as two independent quantum events. In chapter 3, we show quantum gates that superpose, entangle, and detangle qbits using conditional gates.

1.6 Another example of qbit

Other than spins, another example of qbit is linearly polarized photons. Using a linear polarizer, we can obtain the horizontally polarized photon ($\leftrightarrow$) and the vertically polarized photon ($\updownarrow$) as we define $|\leftrightarrow\rangle = |0\rangle = \begin{bmatrix} 1 \\ 0 \end{bmatrix}$ and $|\updownarrow\rangle = |1\rangle = \begin{bmatrix} 0 \\ 1 \end{bmatrix}$, respectively.

A photon with an arbitrary linear polarization can be expressed as $|\psi\rangle = a|\updownarrow\rangle + b|\leftrightarrow\rangle$ because of the orthonormal condition:

$$\langle\updownarrow|\updownarrow\rangle = \langle\leftrightarrow|\leftrightarrow\rangle = 1 \text{ and } \langle\updownarrow|\leftrightarrow\rangle = \langle\leftrightarrow|\updownarrow\rangle = 0. \quad (1.37)$$

For photon qbits, the projection operators represent the observation of light using a polarizer whose polarization axis is either in the horizontal or in the vertical direction.

$$\hat{P}_0 = |\leftrightarrow\rangle\langle\leftrightarrow| = |0\rangle\langle0| = \begin{bmatrix} 1 \\ 0 \end{bmatrix}[1 \ 0] = \begin{bmatrix} 1 & 0 \\ 0 & 0 \end{bmatrix}, \quad (1.38)$$

and

$$\hat{P}_1 = |\updownarrow\rangle\langle\updownarrow| = |1\rangle\langle1| = \begin{bmatrix} 0 \\ 1 \end{bmatrix}[0 \ 1] = \begin{bmatrix} 0 & 0 \\ 0 & 1 \end{bmatrix}. \quad (1.39)$$

Similar to the electron spin, the probability of observing the horizontally polarized photon is given by $|a|^2$ from the measurement

$\hat{P}_0|\psi\rangle = |0\rangle\langle 0|\psi\rangle = |0\rangle\langle 0|(a|0\rangle + b|1\rangle) = a|0\rangle = a\begin{bmatrix} 1 \\ 0 \end{bmatrix}$, and the probability of observing the vertically polarized photon is given by $|b|^2$ from the measurement

$$\hat{P}_1|\psi\rangle = |1\rangle\langle 1|\psi\rangle = |1\rangle\langle 1|(a|0\rangle + b|1\rangle) = b|1\rangle = b\begin{bmatrix} 0 \\ 1 \end{bmatrix}.$$

We use photon based qbits in chapter 6, Quantum Cryptograph.

Reference

[1]   For a general description of quantum mechanics, refer to, for example ed Liboff R L 2003 *Introductory Quantum Mechanics* (Reading, MA: Addison Wesley)

Chapter 2

Universal gates

Digital computers and quantum computers use universal gates with which any gate can be expressed. However, transistor–transistor logic (TTL) gates for the classical digital computers are not applicable to quantum computers. For quantum computers, there is another set of universal gates that can be implemented with qbits. Let us find out what they are.

2.1 Classical universal gates

Current digital computers have been achieving remarkable performances beyond what was expected. It started from series and parallel connections of two ON–OFF switches to implement binary logic states of AND and OR. Since then, we have been creating faster and smaller switches using electric relays, diodes, vacuum tubes, discrete transistors, and finally integrated circuits. The latest semiconductor fabrication technology is really state of the art, but we all take computational advantages without noticing the sophisticated hardware.

The foundation of digital computers are the various binary logic gates that can be constructed from a single logic gate. Such a logic gate is called a universal gate. Because it is easier to manufacture NAND (NOT + AND) gates in the modern semiconductor fabrication technology, we use a NAND gate to show how to make other binary gates. Figure 2.1 shows the basic binary logic gates and equivalent gates constructed by using NAND gates.

Notice that these logic gates are irreversible in the sense that we lose the input information once the inputs are processed by the gates. We lose the input information once it goes through an irreversible logic gate. Richard Feynman pointed out that irreversible gates are inefficient as thermodynamic irreversibility increases entropy [1]. Considering the thermodynamic relationship between entropy

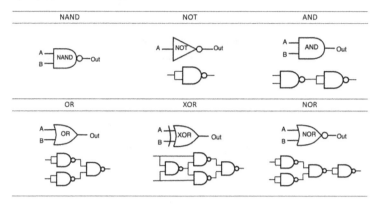

Figure 2.1. Classical binary logic gates.

and heat, $dS = dQ/T$, heat generation is inevitable. For example, a typical super-computer consumes an order of megawatts of electricity where almost all of the electricity is released as heat. Another example of electrical power consumption is with cryptocurrency mining worldwide. Bitcoin mining is consuming more electricity than 159 countries including Ireland and most countries in Africa [2].

It is also known that semiconductor digital electronics is reaching its maximum circuit density according to Moore's law [3]. Quantum computers would solve these issues. We need to 'go quantum' sooner or later for faster computation.

2.2 Alternative universal gates

There is another group of universal gates that includes Control-NOT (CNOT) gate, and Control-control not (CCNOT) gate invented by Tommaso Toffili, and Fredkin gates invented by Edward Fredkin. We show three figures below. Figure 2.2 shows the logic table and the gate diagram of CNOT where an exclusive OR (XOR) is controlled by a single control bit. Figure 2.3 depicts CCNOT where an XOR is controlled by two control bits, figure 2.4 is the Fredkin gate where two bits are to be exchanged by a single control bit. They are reversible because they retain input information, and can be implemented using quantum operators.

Control-NOT gate:

Logic Table				Gate Diagram	
A	**B**	**A'**	**B'**	A ———●———	A'=A
0	0	0	0		
0	1	0	1		
1	0	1	1	B ———⊕———	B'=XOR (A, B)
1	1	1	0		

Figure 2.2. Control-NOT (CNOT) Gate.

Control-control NOT gate:

		Logic Table				Gate Diagram
A	B	C	A'	B'	C'	
0	0	0	0	0	0	
0	0	1	0	0	1	
0	1	0	0	1	0	
0	1	1	0	1	1	
1	0	0	1	0	0	
1	0	1	1	0	1	
1	1	0	1	1	1	
1	1	1	1	1	0	

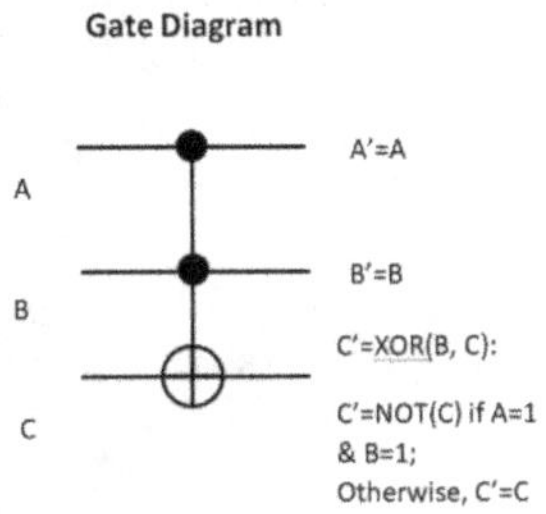

Figure 2.3. Control-Control NOT (CCN) Gate.

		Logic Table				Gate Diagram
A	B	C	A'	B'	C'	
0	0	0	0	0	0	
0	0	1	0	0	1	
0	1	0	0	1	0	
0	1	1	0	1	1	
1	0	0	1	0	0	
1	0	1	1	1	0	
1	1	0	1	0	1	
1	1	1	1	1	1	

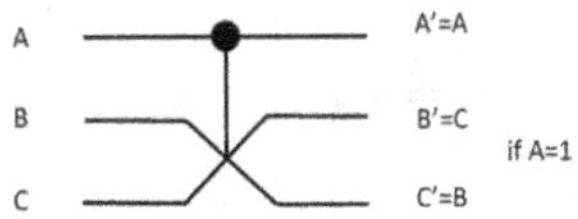

Figure 2.4. Fredkin Gate.

2.3 NOT, CNOT, CCNOT, and Fredkin gates using spin states

If we use the classical binary numbers, these gates do not particularly attract our interest. In order to 'go quantum,' we utilize these gates to create quantum gates. Here, we need to use spin states, |0> and |1>, instead of binary numbers 0 and 1.

2.3.1 NOT-gate

Logic table: A' = NOT(A)   **Gate Diagram**:   A ————●———— A'

Because Pauli's spin matrix, σ_x, reverses the spin orientation (section 1.3.4), we can regard it as a NOT-gate. Then, the matrix representation of the NOT-gate is

$$\hat{U}_{\text{NOT}} = |0\rangle\langle 1| + |1\rangle\langle 0| = \begin{bmatrix} 1 \\ 0 \end{bmatrix}[0 \ 1] + \begin{bmatrix} 0 \\ 1 \end{bmatrix}[1 \ 0] = \begin{bmatrix} 0 & 1 \\ 1 & 0 \end{bmatrix}. \tag{2.1}$$

Operations of the NOT-gate on a single spin state can be described as:

$$\hat{U}_{\text{NOT}}|0\rangle = (|0\rangle\langle 1|+|1\rangle\langle 0|)|0\rangle = \langle 1|0\rangle|0\rangle + \langle 0|0\rangle|1\rangle = \begin{bmatrix} 0 & 1 \\ 1 & 0 \end{bmatrix}\begin{bmatrix} 1 \\ 0 \end{bmatrix} = \begin{bmatrix} 0 \\ 1 \end{bmatrix} = |1\rangle, \tag{2.2}$$

and

$$\hat{U}_{\text{NOT}}|1\rangle = (|0\rangle\langle 1|+|1\rangle\langle 0|)|1\rangle = \langle 1|1\rangle|0\rangle + \langle 0|1\rangle|1\rangle = \begin{bmatrix} 0 & 1 \\ 1 & 0 \end{bmatrix}\begin{bmatrix} 0 \\ 1 \end{bmatrix} = \begin{bmatrix} 1 \\ 0 \end{bmatrix} = |0\rangle$$

where we have used the orthonormal property of the spin states. Notice the NOT-gate is unitary (section 1.2.1).

$$\hat{U}_{\text{NOT}}^{\dagger}\hat{U}_{\text{NOT}} = \begin{bmatrix} 0 & 1 \\ 1 & 0 \end{bmatrix}\begin{bmatrix} 0 & 1 \\ 1 & 0 \end{bmatrix} = \begin{bmatrix} 1 & 0 \\ 0 & 1 \end{bmatrix}.$$

It is interesting to note that Richard Feynman pointed out that the particle creation/annihilation operators can be considered to be the NOT-gate [1].

2.3.2 CNOT-gate

The CNOT-gate is a conditional NOT-gate. It performs
{If $A=|0>$, then $B'=B$} or {If $A=|1>$, then $B'=\text{NOT}(B)$}.

Using the direct product of matrices (equation (1.17)), the CNOT-gate can be formulated as

$$\hat{U}_{\text{CN}} = \hat{P}_{0(A)} \otimes \hat{I}_B + \hat{P}_{1(A)} \otimes \hat{U}_{\text{NOT}(B)}$$
$$= (|0_A\rangle\langle 0_A|) \otimes \hat{I}_B + [|1_A\rangle\langle 1_A| \otimes (|0_B\rangle\langle 1_B| + |1_B\rangle\langle 0_B|)]. \quad (2.3)$$

In equation (2.3), the operator, $\hat{P}_{n(A)}$, is the projection operator for the input A, where $n = 0$ or 1 (equation (1.19)) given by

$$\hat{P}_0 = |0\rangle\langle 0| = \begin{bmatrix} 1 \\ 0 \end{bmatrix}[1\ 0] = \begin{bmatrix} 1 & 0 \\ 0 & 0 \end{bmatrix},\ \hat{P}_1 = |1\rangle\langle 1| = \begin{bmatrix} 0 \\ 1 \end{bmatrix}[0\ 1] = \begin{bmatrix} 0 & 0 \\ 0 & 1 \end{bmatrix}, \quad (2.4)$$

and

$\hat{I}_B = \begin{bmatrix} 1 & 0 \\ 0 & 1 \end{bmatrix}$ is the unit matrix, and $\hat{U}_{\text{NOT}} = |0\rangle\langle 1| + |1\rangle\langle 0| = \begin{bmatrix} 0 & 1 \\ 1 & 0 \end{bmatrix}.$

Operations:
    (a) If $A=|0>$, then output $B'=B$:

$$\hat{U}_{\text{CN}}|0_A\rangle|0_B\rangle = |0_A\rangle\langle 0_A|0_A\rangle\hat{I}_B|0\rangle_B = |0_A\rangle|0\rangle_B,$$
$$\hat{U}_{\text{CN}}|0_A\rangle|1_B\rangle = |0_A\rangle\langle 0_A|0_A\rangle\hat{I}_B|1\rangle_B = |0_A\rangle|1\rangle_B.$$

    (b) If $A=|1>$, then output $B'=\text{NOT}(B)$:

$$\hat{U}_{\text{CN}}|1_A\rangle|0_B\rangle = |1_A\rangle\langle 1_A|1_A\rangle\hat{U}_{\text{NOT}(B)}|0_B\rangle = |1_A\rangle|1_B\rangle,$$
$$\hat{U}_{\text{CN}}|1_A\rangle|1_B\rangle = |1_A\rangle\langle 1_A|1_A\rangle\hat{U}_{\text{NOT}(B)}|1_B\rangle = |1_A\rangle|0_B\rangle.$$

Matrix representation:
    Following the calculation rules of direct products, we obtain

$$|0_A\rangle\langle 0_A|) \otimes \hat{I}_B = \begin{bmatrix} 1 \\ 0 \end{bmatrix}[1\ 0] \otimes \hat{I}_B = \begin{bmatrix} 1 & 0 \\ 0 & 0 \end{bmatrix} \otimes \hat{I}_B = \begin{bmatrix} \hat{I}_B & \hat{0} \\ \hat{0} & \hat{0} \end{bmatrix},$$

$$|1_A\rangle\langle 1_A| \otimes \hat{U}_{\text{NOT}(B)} = \begin{bmatrix} 0 \\ 1 \end{bmatrix}[0\ 1] \otimes \hat{U}_{\text{NOT}(B)} = \begin{bmatrix} 0 & 0 \\ 0 & 1 \end{bmatrix} \otimes \hat{U}_{\text{NOT}(B)} = \begin{bmatrix} \hat{0} & \hat{0} \\ \hat{0} & \hat{U}_{\text{NOT}} \end{bmatrix}$$

where $\hat{O}$ is the 2×2 null matrix. Adding the two gates, we obtain

$$\hat{U}_{\mathrm{CN}} = (|0_A\rangle\langle 0_A|) \otimes \hat{I}_B + |_A\rangle\langle 1_A| \otimes \hat{U}_{\mathrm{NOT}(B)} = \begin{bmatrix} \hat{I}_B & \hat{O} \\ \hat{O} & \hat{U}_{\mathrm{NOT}} \end{bmatrix} = \begin{bmatrix} 1 & 0 & 0 & 0 \\ 0 & 1 & 0 & 0 \\ 0 & 0 & 0 & 1 \\ 0 & 0 & 1 & 0 \end{bmatrix}. \tag{2.5}$$

From the matrix representation, we can easily confirm that the CNOT-gate is unitary: $\hat{U}_{\mathrm{CN}}\hat{U}_{\mathrm{CN}}^{\dagger} = \hat{I}$. The CNOT-gate is very useful to construct quantum gates. We will use the gate in the next chapters to construct quantum circuits.

2.3.3 CCNOT-gate (Toffoli gate)

The CCNOT-gate is a NOT-gate with two control conditions. The logical flow of the CCNOT-gate is
  {If $A=|0>$ & $B=|0>$, $C'=C$} or {If $A=|0>$ & $B=|1>$, $C'=C$},
  or
  {If $A=|1>$ & $B=|0>$, $C'=C$} or {If $A=|1>$ & $B=|1>$, $C'=$NOT C}.

Using the spin notation, the CCNOT-gate can be expressed as

$$\begin{aligned}
\hat{U}_{\mathrm{CCNOT}} &= \hat{P}_{0(A)} \otimes \hat{P}_{0(B)} \otimes \hat{I}_C + \hat{P}_{0(A)} \otimes \hat{P}_{1(B)} \otimes \hat{I}_C \\
&\quad + \hat{P}_{1(A)} \otimes \hat{P}_{0(B)} \otimes \hat{I}_C + \hat{P}_{1(A)} \otimes \hat{P}_{1(B)} \otimes \hat{U}_{\mathrm{NOT}(C)} \\
&= (|0_A\rangle\langle 0_A|) \otimes (|0_B\rangle\langle 0_B|) \otimes \hat{I}_C + (|0_A\rangle\langle 0_A|) \otimes (|1_B\rangle\langle 1_B|) \\
&\quad \otimes \hat{I}_C + (|1_A\rangle\langle 1_A|) \otimes (|0_B\rangle\langle 0_B|) \otimes \hat{I}_C \\
&\quad + (|1_A\rangle\langle 1_A|) \otimes \{(|0_B\rangle\langle 0_B|) \otimes (|0_C\rangle\langle 0_C|) + (|1_B\rangle\langle 1_B|) \otimes (|1_C\rangle\langle 1_C|)\} \\
&\quad + (|1_A\rangle\langle 1_A|) \otimes \{(|0_B\rangle\langle 1_B|) \otimes (|1_C\rangle\langle 0_C|) + (|1_B\rangle\langle 0_B|) \otimes (|0_C\rangle\langle 1_C|)\}.
\end{aligned} \tag{2.6}$$

Operations:
  (a) If $A=|1>$ and $B=|0>$, then $C'=C$:
    $\hat{U}_{\mathrm{CCNOT}}(|0_A\rangle|1_B\rangle|1_C\rangle) = |0_A\rangle|1_B\rangle|1_C\rangle$, and $\hat{U}_{\mathrm{CCNOT}}(|1_A\rangle|0_B\rangle|1_C\rangle) = |1_A\rangle|0_B\rangle|1_C\rangle$.
  (b) If $A=|0>$ and $B=|1>$, then $C'=C$:
    $\hat{U}_{\mathrm{CCNOT}}(|0_A\rangle|1_B\rangle|0_C\rangle) = |0_A\rangle|1_B\rangle|0_C\rangle$, and $\hat{U}_{\mathrm{CCNOT}}(|0_A\rangle|1_B\rangle|1_C\rangle) = |0_A\rangle|1_B\rangle|1_C\rangle$.
  (c) If $A=|0>$ and $B=|0>$, then $C'=C$:
    $\hat{U}_{\mathrm{CCNOT}}(|0_A\rangle|0_B\rangle|0_C\rangle) = |0_A\rangle|0_B\rangle|0_C\rangle$, and $\hat{U}_{\mathrm{CCNOT}}(|0_A\rangle|0_B\rangle|1_C\rangle) = |0_A\rangle|0_B\rangle|1_C\rangle$.
  (d) If $A=|1>$ and $B=|1>$, then output $C'=$NOT(C):
    $\hat{U}_{\mathrm{CCNOT}}(|1_A\rangle|1_B\rangle|0_C\rangle) = |1_A\rangle|1_B\rangle|1_C\rangle$, and $\hat{U}_{CCNOT}(|1_A\rangle|1_B\rangle|1_C\rangle) = |1_A\rangle|1_B\rangle|0_C\rangle$.

Matrix representation:
Similar to the CNOT-gate, we follow the direct product calculation to obtain the matrix representation of the CCNOT-gate.

$$(|0_A\rangle\langle0_A|) \otimes (|0_B\rangle\langle0_B|) \otimes \hat{I}_C = \begin{bmatrix} 1 & 0 & \hat{O} \\ 0 & 0 & \\ \hat{O} & \hat{O} \end{bmatrix} \otimes \hat{I}_C = \begin{bmatrix} \hat{I} & \hat{O} & \hat{O} & \hat{O} \\ \hat{O} & \hat{O} & \hat{O} & \hat{O} \\ \hat{O} & \hat{O} & \hat{O} & \hat{O} \\ \hat{O} & \hat{O} & \hat{O} & \hat{O} \end{bmatrix},$$

and

$$(|0_A\rangle\langle0_A|) \otimes (|1_B\rangle\langle1_B|) \otimes \hat{I}_C = \begin{bmatrix} \hat{O} & \hat{O} & \hat{O} & \hat{O} \\ \hat{O} & \hat{I} & \hat{O} & \hat{O} \\ \hat{O} & \hat{O} & \hat{O} & \hat{O} \\ \hat{O} & \hat{O} & \hat{O} & \hat{O} \end{bmatrix}$$

where $\hat{O}$ is the 2×2 null matrix, and $\hat{I}$ is the 2×2 unit matrix. The remaining gates can be expressed as

$$(|1_A\rangle\langle1_A|) \otimes (|0_B\rangle\langle0_B|) \otimes \hat{I}_C = \begin{bmatrix} \hat{O} & \hat{O} & \hat{O} & \hat{O} \\ \hat{O} & \hat{O} & \hat{O} & \hat{O} \\ \hat{O} & \hat{O} & \hat{I} & \hat{O} \\ \hat{O} & \hat{O} & \hat{O} & \hat{O} \end{bmatrix},$$

and

$$(|1_A\rangle\langle1_A|) \otimes (|1_B\rangle\langle1_B|) \otimes \hat{U}_{NOT(C)} = \begin{bmatrix} \hat{O} & \hat{O} & \hat{O} & \hat{O} \\ \hat{O} & \hat{O} & \hat{O} & \hat{O} \\ \hat{O} & \hat{O} & \hat{O} & \hat{O} \\ \hat{O} & \hat{O} & \hat{O} & \hat{U}_{NOT} \end{bmatrix}.$$

where the 2×2 NOT-gate is given by equation (2.1). By adding these matrices, we obtain

$$\hat{U}_{CCNOT} = \begin{bmatrix} \hat{I} & \hat{O} & \hat{O} & \hat{O} \\ \hat{O} & \hat{I} & \hat{O} & \hat{O} \\ \hat{O} & \hat{O} & \hat{I} & \hat{O} \\ \hat{O} & \hat{O} & \hat{O} & \hat{U}_{NOT} \end{bmatrix}. \tag{2.7}$$

Notice the Toffoli gate is also unitary: $\hat{U}_{CCNOT}\hat{U}^{\dagger}_{CCNOT} = \hat{I}$.

2.3.4 Fredkin gate

The Fredkin gate is a conditional exchange (swap) gate. Its logical statement is given by
{If $A=|0>$, then $B'=B$ & $C'=C$} or {If $A=|1>$, then $B'=C$ & $C' = B$},

which can be formulated as

$$
\begin{aligned}
\hat{U}_{FK} &= \hat{P}_{0(A)} \otimes \hat{I}_B \otimes \hat{I}_C + \hat{P}_{1(A)} \otimes \hat{U}_{EX(BC)} \\
&= |0_A\rangle\langle 0_A| \otimes (|0_B\rangle\langle 0_B| + |1_B\rangle\langle 1_B|) \otimes (|0_C\rangle\langle 0_C| + |1_C\rangle\langle 1_C|) \\
&\quad + |1_A\rangle\langle 1_A| \otimes |0_B\rangle\langle 0_B| \otimes |0_C\rangle\langle 0_C| \\
&\quad + |1_A\rangle\langle 1_A| \otimes [(|0_B\rangle\langle 1_B|) \otimes (|1_C\rangle\langle 0_C|) \\
&\quad + (|1_B\rangle\langle 0_B|) \otimes (|0_C\rangle\langle 1_C|) + (|1_B\rangle\langle 1_B|) \otimes (|1_C\rangle\langle 1_C|)].
\end{aligned}
\tag{2.8}
$$

The exchange gate, $\hat{U}_{EX(BC)}$, is somewhat complicated but it can be confirmed below.

Operations
(1) If $A=|0\rangle$, B and C remain the same states. The Fredkin gate becomes $\hat{U}_{FK} = \hat{P}_{(0)A} \otimes \hat{I}_B \otimes \hat{I}_C$ to output the statement.
(2) If A=$|1\rangle$ and B=$|0\rangle$ and C=$|0\rangle$.

$$
\begin{aligned}
\hat{U}_{FK}|1\rangle_A|0\rangle_B|0\rangle_C &= [(|0_B\rangle\langle 0_B|)(|0_C\rangle\langle 0_C|) + (|0_B\rangle\langle 1_B|)(|1_C\rangle\langle 0_C|) \\
&\quad + (|1_B\rangle\langle 0_B|)(|0_C\rangle\langle 1_C|) + (|1_B\rangle\langle 1_B|)(|1_C\rangle\langle 1_C|)]|1\rangle_A|0\rangle_B|0\rangle_C \\
&= (|0_B\rangle\langle 0_B|)(|0_C\rangle\langle 0_C|)|0\rangle_B|0\rangle_C = |1\rangle_A|0\rangle_B|0\rangle_C.
\end{aligned}
$$

(3) If $A=|1\rangle$ and $B=|1\rangle$ and $C=|0\rangle$.

$$
\begin{aligned}
\hat{U}_{FK}|1\rangle_A|1\rangle_B|0\rangle_C &= [(|0_B\rangle\langle 0_B|)(|0_C\rangle\langle 0_C|) + (|0_B\rangle\langle 1_B|)(|1_C\rangle\langle 0_C|) \\
&\quad + (|1_B\rangle\langle 0_B|)(|0_C\rangle\langle 1_C|) + (|1_B\rangle\langle 1_B|)(|1_C\rangle\langle 1_C|)]|1\rangle_A|1\rangle_B|0\rangle_C \\
&= (|0_B\rangle\langle 1_B|)(|1_C\rangle\langle 0_C|)|1\rangle_B|0\rangle_C = |1\rangle_A|0\rangle_B|1\rangle_C.
\end{aligned}
$$

(4) If $A=|1\rangle$ and $B=|0\rangle$ and $C=|1\rangle$.

$$
\begin{aligned}
\hat{U}_{FK}|1\rangle_A|0\rangle_B|1\rangle_C &= [(|0_B\rangle\langle 0_B|)(|0_C\rangle\langle 0_C|) + (|0_B\rangle\langle 1_B|)(|1_C\rangle\langle 0_C|) \\
&\quad + (|1_B\rangle\langle 0_B|)(|0_C\rangle\langle 1_C|) + (|1_B\rangle\langle 1_B|)(|1_C\rangle\langle 1_C|)]|1\rangle_A|0\rangle_B|1\rangle_C \\
&= (|1_B\rangle\langle 0_B|)(|0_C\rangle\langle 1_C|)|0\rangle_B|1\rangle_C = |1\rangle_A|1\rangle_B|0\rangle_C.
\end{aligned}
$$

(5) If $A=|1\rangle$ and $B=|1\rangle$ and $C=|1\rangle$.

$$
\begin{aligned}
\hat{U}_{FK}|1\rangle_A|1\rangle_B|1\rangle_C &= [(|0_B\rangle\langle 0_B|)(|0_C\rangle\langle 0_C|) + (|0_B\rangle\langle 1_B|)(|1_C\rangle\langle 0_C|) \\
&\quad + (|1_B\rangle\langle 0_B|)(|0_C\rangle\langle 1_C|) + (|1_B\rangle\langle 1_B|)(|1_C\rangle\langle 1_C|)]|1\rangle_A|1\rangle_B|1\rangle_C \\
&= (|1_B\rangle\langle 1_B|)(|1_C\rangle\langle 1_C|)|1\rangle_B|1\rangle_C = |1\rangle_A|1\rangle_B|1\rangle_C.
\end{aligned}
$$

Matrix representation:
The direct product calculations in the Fredkin gate are very similar to the CNOT-gate and the CCNOT-gate.

$$\hat{U}_{FK} = \begin{bmatrix} \hat{I} & \hat{O} & \hat{O} & \hat{O} \\ \hat{O} & \hat{O} & \hat{O} & \hat{O} \\ \hat{O} & \hat{O} & \hat{O} & \hat{O} \\ \hat{O} & \hat{O} & \hat{O} & \hat{O} \end{bmatrix} + \begin{bmatrix} \hat{O} & \hat{O} & \hat{O} & \hat{O} \\ \hat{O} & \hat{I} & \hat{O} & \hat{O} \\ \hat{O} & \hat{O} & \hat{O} & \hat{O} \\ \hat{O} & \hat{O} & \hat{O} & \hat{O} \end{bmatrix} + \begin{bmatrix} \hat{O} & \hat{O} & \hat{O} & \hat{O} \\ \hat{O} & \hat{O} & \hat{O} & \hat{O} \\ \hat{O} & \hat{O} & \hat{I} & \hat{O} \\ \hat{O} & \hat{O} & \hat{O} & \hat{O} \end{bmatrix}$$

$$+ \begin{bmatrix} \hat{O} & \hat{O} & \hat{O} & \hat{O} \\ \hat{O} & \hat{O} & \hat{O} & \hat{O} \\ \hat{O} & \hat{O} & \hat{O} & \begin{smallmatrix}0&0\\1&0\end{smallmatrix} \\ \hat{O} & \hat{O} & \hat{O} & \hat{O} \end{bmatrix} + \begin{bmatrix} \hat{O} & \hat{O} & \hat{O} & \hat{O} \\ \hat{O} & \hat{O} & \hat{O} & \hat{O} \\ \hat{O} & \hat{O} & \hat{O} & \hat{O} \\ \hat{O} & \hat{O} & \begin{smallmatrix}0&1\\0&0\end{smallmatrix} & \hat{O} \end{bmatrix} + \begin{bmatrix} \hat{O} & \hat{O} & \hat{O} & \hat{O} \\ \hat{O} & \hat{O} & \hat{O} & \hat{O} \\ \hat{O} & \hat{O} & \hat{O} & \hat{O} \\ \hat{O} & \hat{O} & \hat{O} & \begin{smallmatrix}0&0\\0&1\end{smallmatrix} \end{bmatrix} \quad (2.9)$$

$$= \begin{bmatrix} \hat{I} & \hat{O} & \hat{O} & \hat{O} \\ \hat{O} & \hat{I} & \hat{O} & \hat{O} \\ \hat{O} & \hat{O} & \hat{I} & \begin{smallmatrix}0&0\\1&0\end{smallmatrix} \\ \hat{O} & \hat{O} & \begin{smallmatrix}0&1\\0&0\end{smallmatrix} & \begin{smallmatrix}0&0\\0&1\end{smallmatrix} \end{bmatrix}.$$

The unitary property of the Fredkin gate can be shown from its matrix representation (2.9).

In this chapter, we did not take full advantage of the quantum states because we simply replaced 0 and 1 with |0> and |1> without using qbits. However, we confirmed how spin states change after these gates, demonstrating that these universal gates can be incorporated with spin states. Notice that the CNOT-gate can incorporate two spin states in certain conditions. Furthermore, if we use the superposed many pairs of qbits, the calculation speed can be dramatically accelerated (section 1.5.2). In chapter 3, we introduce various quantum gates that can be applied to qbits and follow their spin dynamics.

References

[1] Feynman R P 1996 *Feynman Lectures on Computation* (Cambridge: Perseus Publishing)
[2] Lu M 2021 *Visualizing the Power Consumption of Bitcoin Mining* (www.visualcapitalist.com)
[3] Intel 2021 *Over 50 Years of Moore's Law* (www.intel.com)

Chapter 3

Quantum logic gates

We are now ready to investigate quantum gates. They use Pauli's spin matrices described in chapter 1 and the universal gates, the CNOT-gate in particular, described in chapter 2, and other spin operators. In order to be familiar with quantum computation, it is important to trace how qbits change after quantum gates although qbits themselves are not observable. In this chapter, we follow the qbit dynamics in quantum gates step by step.

3.1 Introduction to quantum gate simulation—Blueqat for Python

There are several software programs that simulate quantum gates [1–3]. These simulation programs also offer many examples and explanations of quantum algorithms on their websites, and readers should take advantage of these free resources. Blueqat for Python is one of easy-to-use quantum gate simulators available to everyone, and we use it to demonstrate quantum gates and algorithms. Before we explain quantum gates, this section demonstrates how to install Python and Blueqat on a computer, and how to simulate the quantum gates. No knowledge of Python programming is required to run Blueqat.

3.1.1 Installation of Python and Blueqat

As of May 2021, the version of Python that works with Blueqat is Version 3.7.2 (either the 64-bit version or 32-bit version). Download it from the Python website, Download Python | Python.org.

You may also install Integrated Development and Learning Environment (IDLE) for Python [4], which is more convenient to write programs. For a quick test, type **print ("Hello")** to confirm it is displayed on IDLE as shown in figure 3.1.

To install Blueqat, the default command-line interpreter of Microsoft Windows, cmd. exe, is used. Referring to figure 3.2, click on the [Search] icon in the taskbar, and type

```
Python 3.7.2 (tags/v3.7.2:9a3ffc0492, Dec 23 2018, 23:09:28) [MSC v.1916 64 bit (AMD64)] on win32
Type "help", "copyright", "credits" or "license()" for more information.
>>> print("Hello")
Hello
>>>
```

Figure 3.1. Installation of Python.

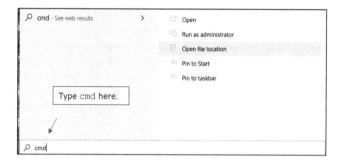

Figure 3.2. Opening command.exe.

```
Microsoft Windows [Version 10.0.19044.1503]
(c) Microsoft Corporation. All rights reserved.

C:\Users\Sammy>pip install blueqat
```
NOTE: The actual screen displays white characters with black background.

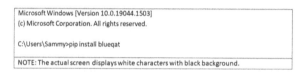

Figure 3.3. Installing Blueqat.

`cmd`, and the command prompt appears. Then type **`pip install blueqat`** and then press `<Enter>` after the command prompt. This step is shown in figure 3.3. Your computer will search for Blueqat from the Internet and install it. Two Python's science libraries, `numpy` and `scipy` are also installed along with Blueqat. Once the installation is successful, a message '`Successfully installed blueqat -x.x.x numpy-x.x.x scipy-x.x.x`' displays. Now we are ready to use Blueqat but remember the following rules.

Remark: Although we do not use Wildqat, which is a quantum annealing simulation, in this book, it can be installed by typing **`pip install wildqat`** on the `cmd` screen.

*** **IMPORTANT** ***

(1) Quantum states are often represented as binary and decimal numbers. For example, four possible states of two superposed qbits are: $|0>|0>=|00>=|0>$ (decimal), $|0>|1>=|01>=|1>$ (decimal), $|1>|0>=|10>=|2>$ (decimal), and $|1>|1>=|11>=|3>$ (decimal). Incorporating this binary to decimal conversion rule and vice versa, quantum circuit diagrams written in this book have *the top qbit line number as the highest quantum bit, i.e., the most significant qbit in the binary number notation.* We use the suffixes of the quantum bit line numbers, e.g., $|\psi>=|0_1>|0_0>$, if necessary.

(2) Blueqat outputs quantum states from 'the smallest decimal-converted state' to the largest. For a quantum circuit of three qbits, the output of the possible

quantum states will be in the order of $|0_2\rangle|0_1\rangle|0_0\rangle=|000\rangle=|0\rangle$ (decimal), $|0_2\rangle|0_1\rangle|1_0\rangle=|001\rangle=|1\rangle$ (decimal), $|0_2\rangle|1_1\rangle|0_0\rangle=|010\rangle=|2\rangle$ (decimal), $|0_2\rangle|1_1\rangle|1_0\rangle=|011\rangle=|3\rangle$ (decimal), $|1_2\rangle|0_1\rangle|0_0\rangle=|100\rangle=|4\rangle$ (decimal), $|1_2\rangle|0_1\rangle|1_0\rangle=|101\rangle=|5\rangle$ (decimal), $|1_2\rangle|1_1\rangle|0_0\rangle=|110\rangle=|6\rangle$ (decimal), and $|1_2\rangle|1_1\rangle|1_0\rangle=|111\rangle=|7\rangle$ (decimal).

(3) When you make observations to find the probabilities of possible quantum states from Blueqat, the first state displayed the **Counter** statement (section 3.2.1) will be *the one with the highest probability* after completing measurements. The numerical values of the resultant probabilities fluctuate due to the stochastic nature.

3.2 Quantum gates

First, we introduce several fundamental single qbit quantum gates with which a single spin is reoriented. Next, we introduce two-qbit quantum gates to create superposition and entanglement of quantum states. We explain frequently-used quantum gates selected for this book. Readers should refer to the Blueqat Tutorial site (GitHub - Blueqat/blueqat-tutorials: Tutorials for Blueqat) for more information on variety of quantum gates.

3.2.1 Pauli's spin matrices

These spin operators rotate a single spin (section 1.3.4). We call them 'quantum gates for a single qbit' in the quantum computation/information field.

$$\text{Pauli X-gate: } \hat{\sigma}_x|0\rangle = \begin{bmatrix} 0 & 1 \\ 1 & 0 \end{bmatrix}\begin{bmatrix} 1 \\ 0 \end{bmatrix} = \begin{bmatrix} 0 \\ 1 \end{bmatrix} = |1\rangle, \text{ and } \hat{\sigma}_x|1\rangle = \begin{bmatrix} 0 & 1 \\ 1 & 0 \end{bmatrix}\begin{bmatrix} 0 \\ 1 \end{bmatrix} = \begin{bmatrix} 1 \\ 0 \end{bmatrix} = |0\rangle;$$

$$\text{Pauli Y-gate: } \hat{\sigma}_y|0\rangle = \begin{bmatrix} 0 & -i \\ i & 0 \end{bmatrix}\begin{bmatrix} 1 \\ 0 \end{bmatrix} = \begin{bmatrix} i \\ 0 \end{bmatrix} = i|0\rangle \text{ and } \hat{\sigma}_y|1\rangle = \begin{bmatrix} 0 & -i \\ i & 0 \end{bmatrix}\begin{bmatrix} 0 \\ 1 \end{bmatrix} = \begin{bmatrix} -i \\ 0 \end{bmatrix} = -i|0\rangle$$

and

$$\text{Pauli Z-gate: } \hat{\sigma}_z|0\rangle = \begin{bmatrix} 1 & 0 \\ 0 & -1 \end{bmatrix}\begin{bmatrix} 1 \\ 0 \end{bmatrix} = \begin{bmatrix} 1 \\ 0 \end{bmatrix} = |0\rangle, \text{ and } \hat{\sigma}_z|1\rangle = \begin{bmatrix} 1 & 0 \\ 0 & -1 \end{bmatrix}\begin{bmatrix} 0 \\ 1 \end{bmatrix} = \begin{bmatrix} 0 \\ -1 \end{bmatrix} = -|1\rangle.$$

How does Blueqat simulate quantum gates with specific codes? In the diagrams below, qbit 0 or $q[0]$ is the input qbit label, squared characters such as $\hat{X}$, $\hat{Y}$, and $\hat{Z}$ are gate operators, and [output] shows the quantum state after a quantum gate. Notice that the default input qbit is always $|0\rangle$, and if we need $|1\rangle$, we apply $\hat{X}$ to $|0\rangle$ as shown below. We use [output] when we conduct measurements. An example of [output] can be seen in figure 3.4.

Blueqat codes
Open IDLE and type
from blueqat import Circuit
to import Blueqat circuits to Python. We need to write this line for any new program. Once a prompt (>>>) appears, enter the codes in the following format:
Circuit().Operator[input qbit line number].run()
where `Circuit()` calls quantum circuits of Blueqat, and

`.Operator[input qbit line number]` stands for:

`.x[0]` is the X-gate applied to |0> of the input qbit line number 0 ($q[0]$);

`.y[0]` is the Y-gate applied to |0> of the input qbit line number 0 ($q[0]$);

`.z[0]` is the X-gate applied to |0> of the input qbit line number 0 ($q[0]$); and

`.run()` means run the command line to output the resultant quantum state *without observation*.

Note: to measure the probabilities of the possible spin states, use the measurement command, `.m[:].run()`, where `[:]` means measure all qbit lines involved in the circuit. For a single bit operator, `.m[:]` means `.m[0]`. In the command, `.m[:].run()`, the `.run()` command specifies the number of measurements to be carried out with the numerical value written in the parentheses. For example, if we repeat observing the output from a quantum circuit 50 times, then enter

`Circuit().Operator[input qbit line number].m[:].run(shots=50)`

Blueqat outputs

If we do not specify the number of observations and use the `.run()` command, Blueqat returns the resultant quantum states. Figure 3.5 is a screenshot of the X-, Y-,

Pauli gates	Matrix operation	Blueqat code	Quantum gate diagram
X-gate	$\hat{\sigma}_x\,|0\rangle =|1\rangle$	`.x[n]`	q[0]: |0> ——[$\hat{X}$]—— 〚output〛
Y-gate	$\hat{\sigma}_y\,|0\rangle = e^{i\pi/2}\,|0\rangle = i\,|0\rangle$	`.y[n]`	q[0]: |0> ——[$\hat{Y}$]—— 〚output〛
Z-gate	$\hat{\sigma}_{z|}\,|0\rangle =|0\rangle$	`.z[n]`	q[0]: |0> ——[$\hat{Z}$]—— 〚output〛

NOTE: The number, n, in the parentheses in the Blueqat code shown above specifies the input qbit line number, similar to the qbit number $q[0]$.

Figure 3.4. X, Y, and Z gates.

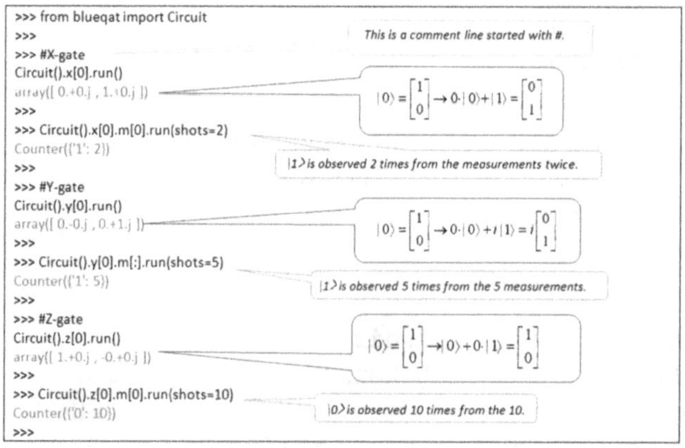

Figure 3.5. Pauli's gate operations.

and Z-codes and their outputs in **array()**. The **array()** statement displays the resultant quantum state vector where the complex unit is denoted as j. For example, **array ([0.+0.j, 1+0.j])** displays the complex coefficients of quantum states |0> and |1>, which are 0 and 1, respectively. Thus, it indicates 0|0>+1|1>=|1>.

In the same figure, the measurement result will be displayed in the **Counter()** statement which shows the number of occurrences out of the total number of measurements specified by the **.run(shots=*)** command. For example, **Counter ({'1': 2})** means that the down spin |1> is observed two times from the two-time measurements as specified by **.m[:].run(shots=2)**.

3.2.2 Hadamard gate (H-gate)

One of the most important quantum gates is the H-gate, which is a a $\pi/2$-rotation around the y-axis, and then a π-rotation around the x-axis, which can be expressed using Pauli's spin matrices, $\hat{\sigma}_x$ and $\hat{\sigma}_z$. Figure 3.6 is the matrix representation of the H-gate.

Blueqat codes and outputs
A measurement after an H-gate should bring both spin states |0> and |1>, which are observed with 'equal' probability. As shown in figure 3.7, for example, the quantum state |1> after applying the H-gate, $\hat{H}|1\rangle = (1/\sqrt{2})(|0\rangle - |1\rangle)$, is displayed as **array**

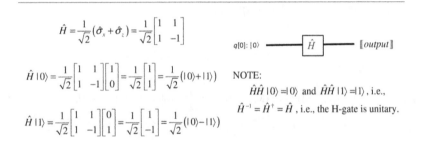

Figure 3.6. Hadamard-gate.

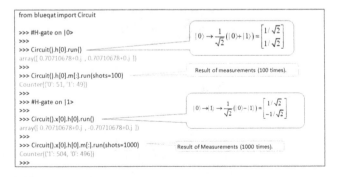

Figure 3.7. H-gate operation.

`([0.70710678+0.j,-0.70710678+0.j])`, and the measurement output, `Counter ({'1': 504; '0': 496})`, which indicates that the down spin |1> is observed 504 times and the up spin |0> is observed 496 times from the measurements repeated 1000 times. Thus, the probabilities are essentially 0.5 each. For this particular measurement, the probability of observing the spin down state was somewhat higher and displayed first.

3.2.3 Superposition of two qbits by applying an H-gate to each qbit

An H-gate can be applied to each of the two qbits to create a superposed quantum state. As we pointed out at the beginning of this chapter, the most significant bit, $q[1]$ is the top line of the initial quantum state, and the input of two qbits is $|0_1\rangle \otimes |0_0\rangle = |0_1\rangle|0_0\rangle = |00\rangle$. Figure 3.8 depicts the quantum circuit for superposing two qbits.

Step-by-step analysis

$$|\psi\rangle_{s0} = |0_1\rangle|0_0\rangle,$$

$$|\psi\rangle_{s1} = \hat{H}_1|0_1\rangle \otimes \hat{H}_0|0_0\rangle = \frac{1}{2}(|0_1\rangle + |1_1\rangle) \otimes (|0_0\rangle + |1_0\rangle)$$

$$= \frac{1}{2}(|0_1\rangle|0_0\rangle + |0_1\rangle|1_0\rangle + |1_1\rangle|0_0\rangle + |1_1\rangle|1_0\rangle) = \frac{1}{2}(|00\rangle + |01\rangle + |10\rangle + |11\rangle).$$

Therefore, all possible superposed states will be output with equal probabilities. Be careful not to change the order of qbits.

Similarly,

$$\hat{H}_1|0_1\rangle \otimes \hat{H}_0|1_0\rangle \rightarrow \frac{1}{2}(|0_1\rangle + |1_1\rangle) \otimes (|0_0\rangle - |1_0\rangle) = \frac{1}{2}(|0_1\rangle|0_0\rangle - |0_1\rangle|1_0\rangle + |1_1\rangle|0_0\rangle - |1_1\rangle|1_1\rangle),$$

$$\hat{H}_1|1_1\rangle \otimes \hat{H}_0|0_0\rangle \rightarrow \frac{1}{2}(|0_1\rangle - |1_1\rangle) \otimes (|0_0\rangle + |1_0\rangle) = \frac{1}{2}(|0_1\rangle|0_0\rangle + |0_1\rangle|1_0\rangle - |1_1\rangle|0_0\rangle - |1_1\rangle|1_1\rangle), \quad (3.1)$$

and

$$\hat{H}_1|1_1\rangle \otimes \hat{H}_0|1_0\rangle \rightarrow \frac{1}{2}(|0_1\rangle - |1_1\rangle) \otimes (|0_0\rangle - |1_0\rangle) = \frac{1}{2}(|0_1\rangle|0_0\rangle - |0_1\rangle|1_0\rangle - |1_1\rangle|0_0\rangle + |1_1\rangle|1_1\rangle).$$

Remark: We often use an abbreviation of the H-gate operation as follows:

$$\hat{H}|x\rangle = \sum_{y=0}^{1} \frac{(-1)^{xy}}{\sqrt{2}}|y\rangle. \quad (3.2)$$

where x and y represent 0 or 1. Readers should confirm that equation (3.2) is equivalent to the results shown in figure 3.6.

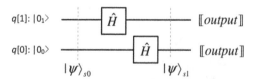

Figure 3.8. Superposition of two qbits using H-gates.

For n-qbits, we denote

$$\hat{H}|x_{n-1}\rangle \otimes \hat{H}|x_{n-2}\rangle \otimes \cdots \otimes \hat{H}|x_0\rangle = \hat{H}^{\otimes n}|x_{n-1}x_{n-2}\cdots x_0\rangle = \hat{H}^{\otimes n}|x\rangle$$

$$= \sum_{y_{n-1}y_{n-2}}\sum \cdots \sum_{y_0}\frac{(-1)^{x_{n-1}y_{n-1}+x_{n-2}y_{n-2}+\cdots+x_0y_0}}{\sqrt{2^n}}|y_{n-1}y_{n-2}\cdots y_0\rangle \qquad (3.3)$$

$$= \sum_{y}\frac{(-1)^{xy}}{\sqrt{2^n}}|y\rangle.$$

where $x \cdot y = x_{n-1}y_{n-1} + x_{n-2}y_{n-2} + \cdots + x_0y_0$, and $|y\rangle = |y_{n-1}y_{n-2}\cdots y_0\rangle$.
When $x = 0$, equation (3.3) becomes

$$\hat{H}^{\otimes n}|0\rangle = \frac{|0\rangle + |1\rangle}{\sqrt{2}} \otimes \frac{|0\rangle + |1\rangle}{\sqrt{2}} \otimes \cdots \otimes \frac{|0\rangle + |1\rangle}{\sqrt{2}}$$

$$= \sum_{y}\frac{1}{\sqrt{2^n}}|y\rangle = \frac{1}{\sqrt{2^n}}(|00\cdots 0\rangle + |00\cdots 1\rangle + \cdots |11\cdots 1\rangle). \qquad (3.4)$$

Blueqat codes and outputs
As we pointed out before, because the default input qbit of Blueqat gates is |0>, if we
need |1>, it must be created by applying the X-gate ($|1\rangle = \hat{X}|0\rangle$). The **array** ()
statements in figure 3.9 display the coeffcients of superposed states |00>, |01>, |10>,
and |11> in this order. The **Counter** () statements are their observed probabilities
from the highest values. The Blueqat correctly displays these outputs shown in
figure 3.9.

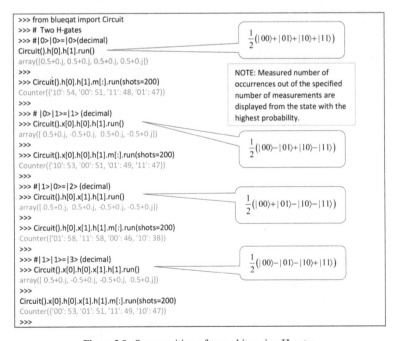

Figure 3.9. Superposition of two qbits using H-gates.

NOTE: From the explicit expressions of |00>, |01>, |10>, and |11> given by equation (1.15), we obtain

$$\frac{1}{2}(|0\rangle|0\rangle + |0\rangle|1\rangle + |1\rangle|0\rangle + |1\rangle|1\rangle) = \frac{1}{2}\begin{bmatrix} 1 \\ 1 \\ 1 \\ 1 \end{bmatrix}, \frac{1}{2}(|0\rangle|0\rangle - |0\rangle|1\rangle + |1\rangle|0\rangle - |1\rangle|1\rangle) = \frac{1}{2}\begin{bmatrix} 1 \\ -1 \\ 1 \\ -1 \end{bmatrix}, \quad (3.5)$$

$$\frac{1}{2}(|0\rangle|0\rangle + |0\rangle|1\rangle - |1\rangle|0\rangle - |1\rangle|1\rangle) = \frac{1}{2}\begin{bmatrix} 1 \\ 1 \\ -1 \\ 1 \end{bmatrix}, \text{ and } \frac{1}{2}(|0\rangle|0\rangle - |0\rangle|1\rangle - |1\rangle|0\rangle - |1\rangle|1\rangle) = \frac{1}{2}\begin{bmatrix} 1 \\ -1 \\ -1 \\ 1 \end{bmatrix}.$$

3.2.4 S-gate and T-gate

These gates represent phase gates ($\pi/2$-rotation and a $\pi/4$-rotation) of a single bit, respectively. Both gates are essentially the same where the rotational angles are $\pi/2$ or $\pi/4$. Figure 3.10 shows their matrix representations and quantum circuits. The corresponding Blueqat codes and outputs are shown in figure 3.11.

3.2.5 Rotational gates at arbitrary angles

The rotational operators are given in chapter 1 (equation (1.12)). Table 3.1 shows the Blueqat codes where θ is the rotational angle at each axis.

Figure 3.12 shows the Blueqat codes from the rotational gates and their outputs where array()'s represent matrix components.

S-gate	T-gate
$\hat{S} = \begin{bmatrix} 1 & 0 \\ 0 & e^{i\frac{\pi}{2}} \end{bmatrix} = \begin{bmatrix} 1 & 0 \\ 0 & i \end{bmatrix}$	$\hat{T} = \begin{bmatrix} 1 & 0 \\ 0 & e^{i\frac{\pi}{4}} \end{bmatrix}$
$\hat{S}^{\dagger} = \begin{bmatrix} 1 & 0 \\ 0 & e^{-i\frac{\pi}{2}} \end{bmatrix} = \begin{bmatrix} 1 & 0 \\ 0 & -i \end{bmatrix}$	$\hat{T}^{\dagger} = \begin{bmatrix} 1 & 0 \\ 0 & e^{-i\frac{\pi}{4}} \end{bmatrix}$

$\hat{S}|0\rangle = |0\rangle$

q[0] ——[$\hat{S}$]—— ⟦output⟧

$\hat{T}|0\rangle = |0\rangle$

q[0] ——[$\hat{T}$]—— ⟦output⟧

$\hat{S}\hat{X}|0\rangle = \hat{S}|1\rangle = e^{i\frac{\pi}{2}}|1\rangle = i|1\rangle$

q[0] ——[$\hat{S}$]——[$\hat{X}$]—— ⟦output⟧

$\hat{T}\hat{X}|0\rangle = \hat{T}|1\rangle = e^{i\frac{\pi}{4}}|1\rangle$

q[0] ——[$\hat{X}$]——[$\hat{T}$]—— ⟦output⟧

NOTE: Blueqat codes for S† and T† are sdg[n] and tdg[n], respectively, where n is the qbit line number.

Figure 3.10. S-gate and T-gate.

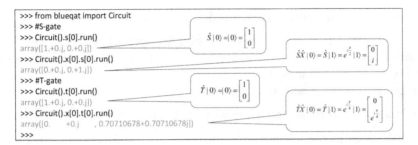

Figure 3.11. S-gate and T-gate.

Table 3.1. Rotational gates about x-, y-, and z-axes at arbitrary angles.

$$\hat{R}_x(\theta) = \left(\cos\frac{\theta}{2}\right)\hat{I} + i\left(\sin\frac{\theta}{2}\right)\hat{\sigma}_x \quad \hat{R}_y(\theta) = \left(\cos\frac{\theta}{2}\right)\hat{I} + i\left(\sin\frac{\theta}{2}\right)\hat{\sigma}_y \quad \hat{R}_z(\theta) = \left(\cos\frac{\theta}{2}\right)\hat{I} + i\left(\sin\frac{\theta}{2}\right)\hat{\sigma}_z$$

Blueqat code: rx (θ)	Blueqat code: ry (θ)	Blueqat code: rz (θ)

Blueqat codes and outputs

In order to use these commands, we need to import a math tool from Python using

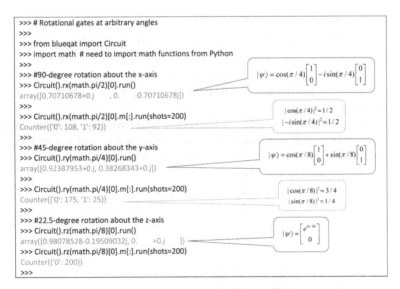

Figure 3.12. Rotational gates at arbitrary angles at around x-, y-, and z-axes.

import math command. For the rotational gates for $\theta = \pi/2$, the angle $\pi/2$ can be calculated by **math.pi/2**. Notice that the spin rotates by angle $\pi/4$ (equations (1.27)–(1.29)) if the specified angle is $\pi/2$ as shown in figure 3.12.

3.3 Controlled-unitary (controlled-U) gates

A controlled gate has at least one control qbit line and a unitary operation. When the control qbit C is $|0>=|\uparrow>$, the target bit is output without any change. When the control bit is $|1>=|\downarrow>$, the target qbit undergoes the unitary transform U. When the target bit is a superposed state, $|\psi>=a|\uparrow>+b|\downarrow>$, the output bit is given by $|\psi'>=a'|\uparrow>+b'|\downarrow>$ where a' and b' are given by the unitary transform: $\begin{bmatrix} a' \\ b' \end{bmatrix} = \hat{U} \begin{bmatrix} a \\ b \end{bmatrix}$.

Figure 3.13 shows the schematic diagram of the controlled-U gate where qbit 0, q [0], is the control bit and qbit 1 is the operated bit, $q[1]$.

3.3.1 CNOT (or CX) gate

The CNOT-gate is a controlled unitary gate as we described in chapter 2. Here, we assign $q[0]$ to be the *control bit* and $q[1]$ to be the *target bit*. If the control bit is $|1>$, then the NOT-gate, which is the X-gate, is applied to the target gate. For this reason, the CNOT-gate is also called the Controlled-X gate or the CX-gate. The CNOT-gate of Blueqat is given by the quantum circuit shown in figure 3.14.

Blueqat codes and outputs

Figure 3.15 shows the Blueqat codes and outputs from the CNOT-gate. Recall the NOTE in section 3.2.2 for the outputs.

3.3.2 Controlled-Z gate and controlled-P gate

Along with the CNOT-gate, the controlled-Z gate (U_{CZ}), and the controlled-phase gate (U_{CP}), are useful later in this book. For the Z-gate and the phase gate, see sections 3.2.1, and 3.2.5. Figure 3.16 shows the quantum logic table and circuit of these gates. Notice the difference between these gates is the unitary gate used in the respective quantum circuit.

The operations of CZ-gate and the CP-gate are shown below. Here, rather than the testing four different combination of two qbits, we create a superposed state, $|00>+|01>+|10>+|11>$, by applying two H-gates (section 3.2.3). This is a major advantage of quantum computation.

NOTE: $|0_1\rangle + |1_1\rangle$ will be also written as $(|0\rangle + |1\rangle)_1$.

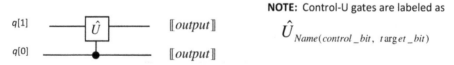

NOTE: Control-U gates are labeled as

$\hat{U}_{Name(control\_bit,\ target\_bit)}$

Figure 3.13. Controlled-unitary gate.

cnot[control bit, target bit] or cx[control bit, target bit]

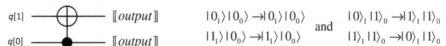

$|0_1\rangle|0_0\rangle \rightarrow |0_1\rangle|0_0\rangle$

$|1_1\rangle|0_0\rangle \rightarrow |1_1\rangle|0_0\rangle$

and

$|0\rangle_1|1\rangle_0 \rightarrow |1\rangle_1|1\rangle_0$

$|1\rangle_1|1\rangle_0 \rightarrow |0\rangle_1|1\rangle_0$

Figure 3.14. CNOT (CX)-gate.

```
>>> # CNOT gate (the control bit is q[0] and the target bit is q[1].
>>> from blueqat import Circuit
>>>
>>> #|target bit>=|0> and |control bit> =|1> :
>>> Circuit().x[0].cnot[0,1].run()
array([0.+0.j, 0.+0.j, 0.+0.j, 1.+0.j])
>>>
>>> Circuit().x[0].cnot[0,1].m[:].run(shots=1000)
Counter({'11': 1000})
>>>
>>> #|target bit>=|1> and |control bit> =|1>
>>> Circuit().x[0].x[1].cx[0,1].run()
array([0.+0.j, 1.+0.j, 0.+0.j, 0.+0.j])
>>>
>>> Circuit().x[0].x[1].cx[0,1].m[:].run(shots=1000)
Counter({'10': 1000})
>>>
```

$|0\rangle|1\rangle \rightarrow |1\rangle|1\rangle$

$|1\rangle|1\rangle \rightarrow |0\rangle|1\rangle$

Figure 3.15. CNOT-gate.

Target: q[1]	Control: q[0]	CZ-gate ($U_{CZ(0,1)}$)	CP-gate ($U_{CP(0,1)}$)
$|0\rangle$	$|0\rangle$	$|0\rangle|0\rangle$	$|0\rangle|0\rangle$
$|1\rangle$	$|0\rangle$	$|1\rangle|0\rangle$	$|1\rangle|0\rangle$
$|0\rangle$	$|1\rangle$	$|0\rangle|1\rangle$	$|0\rangle|1\rangle$
$|1\rangle$	$|1\rangle$	$-|1\rangle|1\rangle$	$e^{i\lambda}|1\rangle|1\rangle$

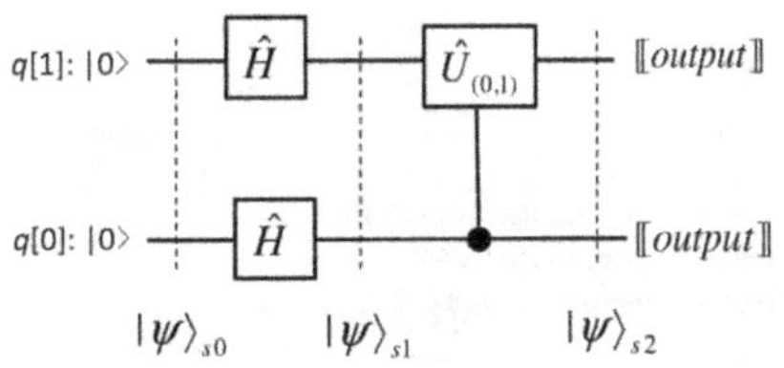

Figure 3.16. CZ-gate and CP-gate.

Step-by-step analyses

$$|\psi\rangle_{s0} = |0_1\rangle|0_0\rangle,$$

$$|\psi\rangle_{s1} = \hat{H}^{\otimes 2}|\psi\rangle_{s0} = \left(\frac{1}{\sqrt{2}}(|0\rangle + |1\rangle)\right)_1\left(\frac{1}{\sqrt{2}}(|0\rangle + |1\rangle)\right)_0.$$

For the CZ-gate,

$$|\psi\rangle_{s2} = \left(\hat{U}_{CZ(0,1)}|\psi\rangle_{s1}\right) = \frac{1}{2}\hat{U}_{CZ(0,1)}((|0\rangle + |1\rangle)_1|0_0\rangle + |(|0\rangle + |1\rangle)_1|1_0\rangle)$$

$$= \frac{1}{2}((|0\rangle + |1\rangle)_1|0_0\rangle + |(|0\rangle - |1\rangle)_1|1_0\rangle) = \frac{1}{2}(|0_1\rangle|0_0\rangle + |0_1\rangle|1_0\rangle + |1_1\rangle|0_0\rangle - |1_1\rangle|1_0\rangle).$$

For the CP-gate with $\lambda=\pi/2$,

$$|\psi\rangle_{s2} = \left(\hat{U}_{CP(0,1)}|\psi\rangle_{s1}\right) = \frac{1}{2}\hat{U}_{CP(0,1)}((|0\rangle + |1\rangle)_1|0_0\rangle + |(|0\rangle + |1\rangle)_1|1_0\rangle)$$

$$= \frac{1}{2}\left((|0\rangle + |1\rangle)_1|0_0\rangle + |(|0\rangle + e^{i\pi/2}|1\rangle)_1|1_0\rangle\right) = \frac{1}{2}(|0_1\rangle|0_0\rangle + |0_1\rangle|1_0\rangle + |1_1\rangle|0_0\rangle + e^{i\pi/2}|1_1\rangle|1_0\rangle).$$

3.3.3 Controlled–Z equivalent circuit

The following gate combinations, the H-U_{CN}-H gate enclosed by the broken line in figure 3.19, is equivalent to the U_{CZ} -gate. Here, the superposed state, |00>+|01> +|10>+|11>, which will be passed through the U_{CZ} equivalent gate to check its equivalence.

Blueqat codes and outputs

Figures 3.17 and 3.18 show the coded and outputs from the CZ gate and the CP gate, respectively.

Step-by-step analysis

$$|\psi\rangle_{s0} = |0_1\rangle|0_0\rangle,$$

$$|\psi\rangle_{s1} = \hat{H}_1\hat{H}_0|\psi\rangle_{s0} = \hat{H}^{\otimes 2}|\psi\rangle_{s0} = \left(\frac{1}{\sqrt{2}}(|0\rangle + |1\rangle)\right)_1\left(\frac{1}{\sqrt{2}}(|0\rangle + |1\rangle)\right)_0,$$

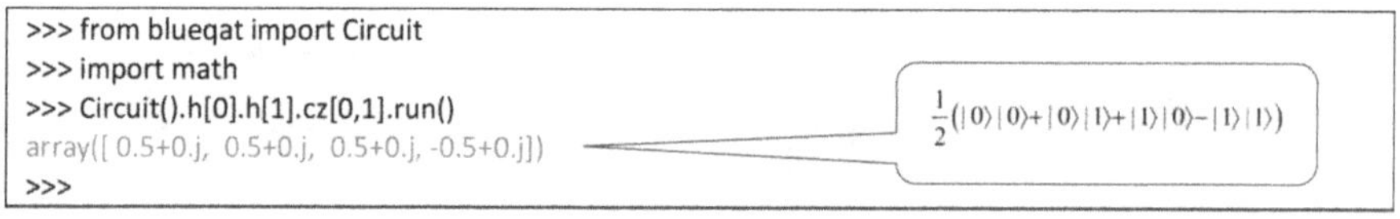

>>> from blueqat import Circuit
>>> import math
>>> Circuit().h[0].h[1].cz[0,1].run()
array([0.5+0.j, 0.5+0.j, 0.5+0.j, -0.5+0.j])
>>>
$\frac{1}{2}(|0\rangle|0\rangle+|0\rangle|1\rangle+|1\rangle|0\rangle-|1\rangle|1\rangle)$

Figure 3.17. Blueqat program of the CZ gate.

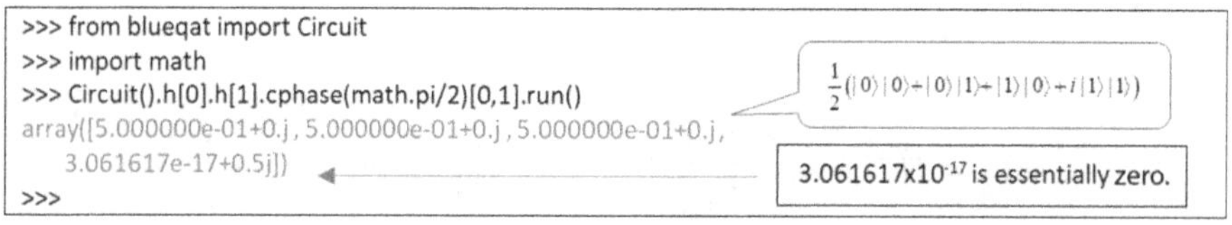

>>> from blueqat import Circuit
>>> import math
>>> Circuit().h[0].h[1].cphase(math.pi/2)[0,1].run()
array([5.000000e-01+0.j, 5.000000e-01+0.j, 5.000000e-01+0.j,
3.061617e-17+0.5j])
>>>
$\frac{1}{2}(|0\rangle|0\rangle+|0\rangle|1\rangle+|1\rangle|0\rangle+i|1\rangle|1\rangle)$
3.061617x10^{-17} is essentially zero.

Figure 3.18. Blueqat program of the CP gate.

q[1]: |0⟩
q[0]: |0⟩
$\hat{H}$
$\hat{H}$
$\hat{H}$
$\hat{H}$
[[output]]
[[output]]
$|\psi\rangle_{s0}$
$|\psi\rangle_{s1}$
$|\psi\rangle_{s2}$
$|\psi\rangle_{s3}$
$|\psi\rangle_{s4}$

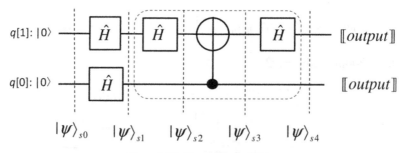

Figure 3.19. U_{CZ} equivalent gate.

But notice

$$|\psi\rangle_{s2} = \left(\hat{H}_1|\psi\rangle_{step1}\right) = \frac{1}{2}\hat{H}_1\hat{H}_1\hat{H}_0|\psi\rangle_{s0} = \frac{1}{2}\hat{H}_0|\psi\rangle_{s0} \text{ because } \hat{H}\hat{H} = \hat{I}.$$

Thus,

$$|\psi\rangle_{s3} = \hat{U}_{CN}|\psi\rangle_{s2} = \frac{1}{\sqrt{2}}\hat{U}_{CN}(|00\rangle + |10\rangle) = \frac{1}{\sqrt{2}}(|00\rangle + |11\rangle),$$

$$|\psi\rangle_{s4} = \hat{H}|\psi\rangle_{s3} = \frac{1}{\sqrt{2}}(|0\rangle\hat{H}|0\rangle + |1\rangle\hat{H}|1\rangle)$$

$$= \frac{1}{\sqrt{2}}\left(|0\rangle\frac{1}{\sqrt{2}}(|0\rangle + |1\rangle) + |1\rangle\frac{1}{\sqrt{2}}(|0\rangle - |1\rangle)\right) = \frac{1}{2}(|00\rangle + |01\rangle + |10\rangle - |11\rangle).$$

Therefore, the H-U_{CN}-H gate is equivalent to the U_{CZ} gate.

3.3.4 SWAP gate

The swap gate can be constructed by three CN-gates as shown figure 3.20.

Step-by-step analysis

$$|\psi\rangle_{s0} = |0_1\rangle|0_0\rangle.$$

(i) Swapping $|1_1\rangle|0_0\rangle$

$$|\psi\rangle_{s1} = \hat{X}_1|0_1\rangle|0_0\rangle = |1_1\rangle|0_0\rangle,$$

$|\psi\rangle_{s2} = \hat{U}_{CN(0,1)}|1_1\rangle|0_0\rangle = |1_1\rangle|0_0\rangle$ after the first CN-gate,
$|\psi\rangle_{s3} = \hat{U}_{CN(1,0)}|1_1\rangle|0_0\rangle = |1_1\rangle|1_0\rangle$ after the second CN-gate,
$|\psi\rangle_{s4} = \hat{U}_{CN(0,1)}|1_1\rangle|1_0\rangle = |0_1\rangle|1_0\rangle$ after the third CN gate.
Thus, $|1\rangle|0\rangle \rightarrow |0\rangle|1\rangle$.

(ii) Swapping $|0_1\rangle|1_0\rangle$.

$$|\psi\rangle_{s1} = \hat{X}_0|0_1\rangle|0_0\rangle = |0_1\rangle|1_0\rangle,$$

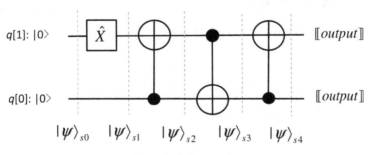

Figure 3.20. Swap gate.

$|\psi\rangle_{s2} = \hat{U}_{CN(0,1)}|0_1\rangle|1_0\rangle = |1_1\rangle|1_0\rangle$ after the first CN-gate,

$|\psi\rangle_{s3} = \hat{U}_{CN(1,0)}|1_1\rangle|1_0\rangle = |1_1\rangle|0_0\rangle$ after the second CN-gate,

$|\psi\rangle_{s4} = \hat{U}_{CN(0,1)}|1_1\rangle|0_0\rangle = |1_1\rangle|0_0\rangle$ after the third CN gate.

Thus, $|0\rangle|1\rangle \rightarrow |1\rangle|0\rangle$.

Blueqat codes and outputs

In both cases, qbits $q[0]$ and $q[1]$ are swapped by the Bueqat codes shown in figure 3.21.

3.3.5 CCNOT gate (Toffoli gate)

There are two controlled bits and one target bit as described in section 2.3.3. If both controlled gates are $|1\rangle$, the target bit is reversed. The quantum circuit of the CCNOT-gate is given by figure 3.22.

Blueqat codes and outputs

In figure 3.23, the X-gate are applied to the control bits, $q[1]$ and $q[0]$, to change the default $|0\rangle$ to $|1\rangle$. Another X-gate is applied to the target bit $q[2]$.

NOTE: The CCNOT-gate can be constructed with the H-gates, the CNOT-gates, and Rz-gates as shown in figure 3.24.

Blueqat codes and outputs

Recall that the Blueqat code of the T^{+}-gate is tdg[n] where n is the q-line number (section 3.2.4). Figure 3.25 shows the quantum circuit equivalent to the CCNOT-gate.

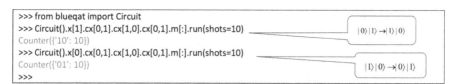

Figure 3.21. Swap-gate.

q[2] ─────⊕───── ⟦*output*⟧

q[1] ─────●───── ⟦*output*⟧

q[0] ─────●───── ⟦*output*⟧

$\hat{U}_{CCX(0,1,2)}$

$|0_2\rangle|1_1\rangle|1_0\rangle \rightarrow |1_2\rangle|1_1\rangle|1_0\rangle$

$|1_2\rangle|1_1\rangle|1_0\rangle \rightarrow |0_2\rangle|1_1\rangle|1_0\rangle$

Otherwise, no change.

Figure 3.22. CCNOT-gate.

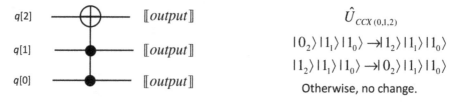

Figure 3.23. Blueqat program of CCNOT-gate.

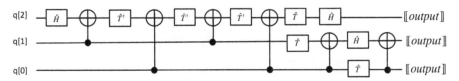

Figure 3.24. CCNOT equivalent gate.

```
>>> from blueqat import Circuit
>>>
Circuit().x[0].x[1].h[2].cnot[1,2].tdg[2].cnot[0,2].t[2].cnot[1,2].tdg[2].cnot[0,2].t[1].t[2].h[2].cnot[0,1].tdg[1]
.t[0].cnot[0,1].m[:].run(shots=50)
Counter({'111': 50})
>>>
Circuit().x[0].x[1].x[2].h[2].cnot[1,2].tdg[2].cnot[0,2].t[2].cnot[1,2].tdg[2].cnot[0,2].t[1].t[2].h[2].cnot[0,1].
tdg[1].t[0].cnot[0,1].m[:].run(shots=50)
Counter({'110': 50})
>>>
```

Figure 3.25. Blueqat program of CCN-gate.

3.3.6 Bell gate

The Bell gate entangles two qbits. The quantum circuit can be constructed by combining an H-gate and a CN-gate as shown in figure 3.26.

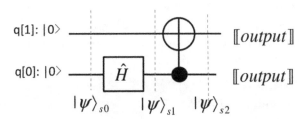

Figure 3.26. Bell gate.

Step-by-step analysis

$$|\psi\rangle_{s0} = |0_1\rangle|0_0\rangle,$$

$$|\psi\rangle_{s1} = |0_1\rangle\hat{H}_0|0_0\rangle = \frac{1}{\sqrt{2}}|0_1\rangle(|0\rangle + |1\rangle)_0 = \frac{1}{\sqrt{2}}(|0_1\rangle|0_0\rangle + |0_1\rangle|1_0\rangle),$$

$$|\psi\rangle_{s2} = \frac{1}{\sqrt{2}}\hat{U}_{CN}(|0_1\rangle|0_0\rangle + |0_1\rangle|1_0\rangle) = \frac{1}{\sqrt{2}}(|0_0\rangle|0_1\rangle + |1_0\rangle|1_1\rangle)$$

$$= \frac{1}{\sqrt{2}}(|00\rangle + |11\rangle)$$

where |00> and |11> are entangled states. Similarly,

$$\hat{B}|0\rangle_1|1\rangle_0 = \frac{1}{\sqrt{2}}(|00\rangle - |11\rangle)), \ \hat{B}|1\rangle_1|0\rangle_0 = \frac{1}{\sqrt{2}}(|10\rangle + |01\rangle)), \text{ and } \hat{B}|1\rangle_1|1\rangle_1$$

$$= \frac{1}{\sqrt{2}}(|10\rangle - |01\rangle))$$

where we can use the X-gate to create |1>.

Blueqat codes and outputs

Figure 3.27 lists the Blueqat codes of the Bell-gate applied to four different input qbits, and entangled outputs. The output displays the complex coefficients of |00>, |01>, |10>, and |11> in this order.

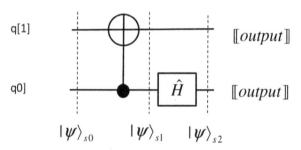

Figure 3.27. Reverse Bell Gate.

3.3.7 Reverse Bell (B^{-1})

The reverse Bell gate detangles two entangled qbits. As shown in figure 3.28, its circuit can be constructed by reversing the operational order of the Bell gates. Notice $\hat{B}\hat{B}^{-1} = \hat{I}$.

Step-by-step analysis

Let start with an entangled qbits.

$$|\psi\rangle_{s0} = \frac{1}{\sqrt{2}}(|0_10_0\rangle + |1_11_0\rangle),$$

$$|\psi\rangle_{s1} = \hat{U}_{CN(0,1)}\frac{1}{\sqrt{2}}(|0_10_0\rangle + |1_11_0\rangle) = \frac{1}{\sqrt{2}}(|0_10_0\rangle + |0_11_0\rangle),$$

$$|\psi\rangle_{s2} = \hat{H}_0\frac{1}{\sqrt{2}}(|0_10_0\rangle + |0_11_0\rangle) = \frac{1}{2}[|0_1\rangle(|0_0\rangle + |1_0\rangle) + |0_1\rangle(|0_0\rangle - |1_0\rangle)] = |0_1\rangle|0_0\rangle.$$

Similarly,

$$\hat{B}^{-1}\left[\frac{1}{\sqrt{2}}(|00\rangle - |11\rangle))\right] = |01\rangle, \ \hat{B}^{-1}\left[\frac{1}{\sqrt{2}}(|10\rangle + |01\rangle))\right] = |10\rangle,$$

$$\hat{B}^{-1}\left[\frac{1}{\sqrt{2}}(|10\rangle - |01\rangle))\right] = |11\rangle.$$

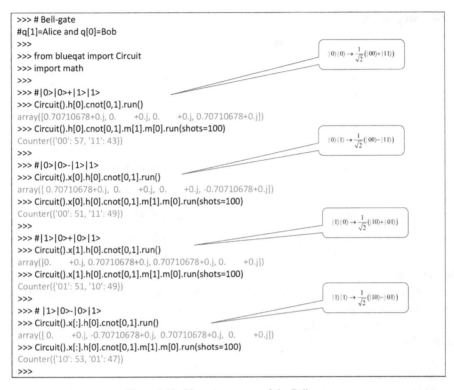

Figure 3.28. Blueqat program of the Bell gate.

As we see, we can use multiple Bell gates to entangle qbits, which can be operated at once for 'parallel computing.' Then we use multiple reverse Bell gates to detangle them to extract the results.

3.3.8 GHZ state

The Greenberger–Horne–Zeilinger (GHZ) state entangles 3 qbits [5]. Figure 3.29 illustrates the quantum circuit of creating the GHZ-state. Figure 3.30 shows its Blueqat operation.

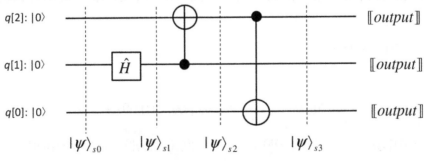

Figure 3.29. GHZ gate.

Step-by step analysis

$$|\psi\rangle_{s0} = |0_2\rangle|0_1\rangle|0_0\rangle,$$

$$|\psi\rangle_{s1} = |0_2\rangle\hat{H}_1|0_1\rangle|0_0\rangle = |0_2\rangle\frac{1}{\sqrt{2}}(|0\rangle + |1\rangle)_1|0_0\rangle$$

$$= \frac{1}{\sqrt{2}}(|0_2\rangle|0_1\rangle|0_2\rangle + |0_2\rangle|1_1\rangle|0_0\rangle),$$

$$|\psi\rangle_{s2} = \hat{U}_{CN(1,2)}|\psi\rangle_{s1} = \frac{1}{\sqrt{2}}(|0_20_1\rangle|0_0\rangle + |1_2\,1_1\rangle|0_0\rangle),$$

$$|\psi\rangle_{s3} = \hat{U}_{CN(2,\,0)}|\psi\rangle_{s2} = \frac{1}{\sqrt{2}}(|0_20_1\rangle0_0 + |1_2\,1_1\,1_0\rangle)$$

$$= \frac{1}{\sqrt{2}}(|000\rangle + |111\rangle).$$

Blueqat codes and outputs
Carefully select the control bit of the CNOT gates, and we obtain the correct output, which show the resultant state of |000>, |001>, |010>, |011>, |100>, |101>, |110>, and |111> in this order where, except |000> and |111>, all are nulls as shown in figure 3.30.

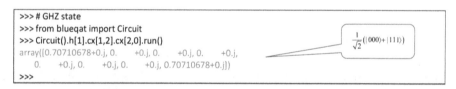

Figure 3.30. Blueqat program of creating GHZ state.

3.4 Half adder from quantum gates

The basic arithmetic gate of classical digital computers is the half adder where there are two inputs, A and B, and the sum $S=XOR(A,B)$ and the carry $C=AND(A,B)$ as shown in figure 3.31, It also shows the truth table.

The quantum half adder shown in figure 3.32 is basically the same as the classical gates. However, the quantum gate can simultaneously calculate 0+0, 0+1, 1+0, and 1+1.

Step-by-step analysis

$$|\psi\rangle_{s0} = |0_3\rangle|0_2\rangle|0_1\rangle|0_0\rangle,$$

$$|\psi\rangle_{s1} = |0_3\rangle|0_2\rangle\hat{H}_1|0_1\rangle\hat{H}_0|0_0\rangle = \frac{1}{2}|0_3\rangle|0_2\rangle(|0\rangle + |1\rangle)_1(|0\rangle + |1\rangle)_0$$

$$= \frac{1}{2}(|0_3\rangle|0_2\rangle|0_1\rangle|0_0\rangle + |0_3\rangle|0_2\rangle|0_1\rangle|1_0\rangle + |0_3\rangle|0_2\rangle|1_1\rangle|0_0\rangle + |0_3\rangle|0_2\rangle|1_1\rangle|1_0\rangle),$$

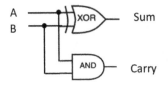

A	B	S	C
0	0	0	0
1	0	1	0
0	1	1	0
1	1	0	1

Figure 3.31. Classical half adder.

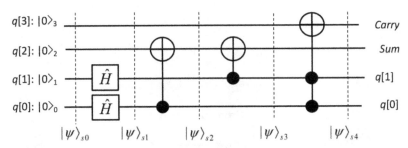

Figure 3.32. Quantum half adder.

$$\left|\psi\right\rangle_{s2} = \hat{U}_{\mathrm{CN}(0,2)}\left|\psi\right\rangle_{s1}$$
$$= \frac{1}{2}(|0_3\rangle|0_2\rangle|0_1\rangle|0_0\rangle + |0_3\rangle|1_2\rangle|0_1\rangle|1_0\rangle + |0_3\rangle|0_2\rangle|1_1\rangle|0_0\rangle + |0_3\rangle|1_2\rangle|1_1\rangle|1_0\rangle),$$

$$\left|\psi\right\rangle_{s3} = \hat{U}_{\mathrm{CN}(1,2)}\left|\psi\right\rangle_{s2}$$
$$= \frac{1}{2}(|0_3\rangle|0_2\rangle|0_1\rangle|0_0\rangle + |0_3\rangle|1_2\rangle|0_1\rangle|1_0\rangle + |0_3\rangle|1_2\rangle|1_1\rangle|0_0\rangle + |0_3\rangle|0_2\rangle|1_1\rangle|1_0\rangle),$$

$$\left|\psi\right\rangle_{s4} = \hat{U}_{\mathrm{CCN}(0,1,3)}\left|\psi\right\rangle_{s3}$$
$$= \frac{1}{2}(|0_3\rangle|0_2\rangle|0_1\rangle|0_0\rangle + |0_3\rangle|1_2\rangle|0_1\rangle|1_0\rangle + |0_3\rangle|1_2\rangle|1_1\rangle|0_0\rangle + |1_3\rangle|0_2\rangle|1_1\rangle|1_0\rangle).$$

Blueqat codes and outputs

The output format is $|(\mathtt{Carry})(\mathtt{Sum})(\mathtt{q[1]})(\mathtt{q[0]})\rangle$. The resultant quantum state is listed in the order of $|0000\rangle$, $|0001\rangle$, ..., $|1111\rangle$. Readers should confirm that the non-zero states are: $|0000\rangle$, $|0101\rangle$, $|0110\rangle$, and $|1011\rangle$.

Note: As we pointed out before, in figure 3.33, the measurement displayed is displayed from the highest probability, which will be for each measurement.

Remark: If calculations of 0+0, 0+1, 1+0, and 1+1 are performed separately, the half adder will be defined and used repeatedly.

```
>>> # Half-adder
>>> from blueqat import Circuit
>>>
>>> Circuit().h[0].h[1].cx[0,2].cx[1,2].ccx[0,1,3].run()
array([0.5+0.j, 0. +0.j, 0. +0.j, 0. +0.j, 0. +0.j, 0.5+0.j, 0.5+0.j,
    0. +0.j, 0. +0.j, 0. +0.j, 0. +0.j, 0.5+0.j, 0. +0.j, 0. +0.j,
    0. +0.j, 0. +0.j])
>>>
>>> Circuit().h[0].h[1].cx[0,2].cx[1,2].ccx[0,1,3].m[:].run(shots=1000)
Counter({'1010': 272, '0000': 258, '1101': 240, '0110': 230})
>>>
```

Figure 3.33. Simultaneous calculation of binary numbers 0+0, 0+1, 1+0, and 1+1.

```
From blueqat import Circuit
Adder=Circuit().ccx[0,1,2].cx[0,3].cx[1,3]
#|0>+|0>:
(Circuit()+adder()).m[:].run(shots=200)
#|0>+|1>:
(Circuit().x[1]+adder()).m[:].run(shots=200)
#|1>+|0>:
(Circuit().x[0]+adder()).m[:].run(shots=200)
#|1>+|1>:
(Circuit().x[0,1]+adder()).m[:].run(shots=200)
```

We now have various quantum gates. In the next chapter, we combine them to build various algorithms.

References

[1] Blueqat *Github—Thmp/Quantum: Quantum COMPUTER Simulation in Python* (https://github.com/thmp/quantum)
[2] Quantum Toolbox in Python *QuTiP—Quantum Toolbox in Python* (https://qutip.org/)
[3] Open-Spruce Quantum Development *Qiskit* (https://qiskit.org/?language=en)
[4] *IDLE—Python 3.9.1 Documentation* (https://www.python.org/downloads/release/python-391/)
[5] *Greenberger-Horne-Zeilinger State*, PHYSICS 419 Lecture note, Spring 2021, University of Illinois (https://courses.physics.illinois.edu/phys419/sp2021/ghztutorial.pdf)

IOP Publishing

Quantum Computation and Quantum Information
Simulation using Python
A gentle introduction
Shinil Cho

Chapter 4

Algorithms of quantum computation

Using the quantum gates introduced in chapter 3, we now see how they are combined to create quantum circuits to run quantum algorithms. In this book, we focus on well-known algorithms, Deutsch's, Grover's, Quantum Fourier Transform (QFT), Shore's, and Simon's. These are the pioneers of stimulating quantum computation.

We also describe corrections of bit flip and phase flip errors in this chapter. The error corrections are critical for implanting practical quantum computers. There are many textbooks [1–3], and articles [4] on quantum computation as well as information available on the Internet. Readers should refer to these sources for more advanced treatments.

4.1 Deutsch's algorithm

This algorithm demonstrated for the first time that the quantum algorithm is faster than the classical [5]. Suppose there are four binary-functions, f_0, f_1, f_2, and f_3 for binary inputs are 0 and 1:

$$f_0(0) = f_0(1) = 0; f_1(0) = 0 \& f_1(1) = 1; f_2(1) = 1 \& f_2(1) = 0; \text{ and } f_3(0) = f_3(1) = 1$$

where f_0 and f_3 are called constant functions, and f_1 and f_2, are called balanced functions. The constant function means that f always outputs the same bit, i.e., $f(0) = f(1)$ whereas the balanced function means that f outputs different bits on different inputs, i.e., $f(0) \neq f(1)$.

The question is:

How many times do you need to evaluate $f_i(x)$ in order to determine if $f_i(x)$ is either constant (f_0 or f_3) or balance (f_1 or f_2) when one of the four functions is given randomly?

Classically, we need to evaluate $f_i(x)$ by substituting 0 and 1 to answer the question:
if $f_i(0) = 0$, then $f_i(x)$ is f_0 or f_1, and if $f_i(1) = 0$, then $f_i(x) = f_0$, or if $f_i(1) = 1$, then $f_i(x) = f_1$; and
if $f_i(0) = 1$, then $f_i(x)$ is f_2 or f_3, and if $f_i(1) = 0$, then $f_i(x) = f_2$ or if $f_i(1) = 1$, then $f_i(x) = f_2$.

Davide Deutsch proposed the algorithm for this task using the following 2-qbit circuit (an oracle). Figure 4.1 shows the output from the oracle.

If we use |0> and |1> separately as the input, then each gate must be used twice to evaluate $f_i(x)$ if it is constant or balanced. Table 4.1 lists all possibilities for this classical approach.

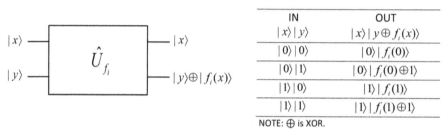

IN	OUT
$\lvert x\rangle\,\lvert y\rangle$	$\lvert x\rangle\,\lvert y\oplus f_i(x)\rangle$
$\lvert 0\rangle\,\lvert 0\rangle$	$\lvert 0\rangle\,\lvert f_i(0)\rangle$
$\lvert 0\rangle\,\lvert 1\rangle$	$\lvert 0\rangle\,\lvert f_i(0)\oplus 1\rangle$
$\lvert 1\rangle\,\lvert 0\rangle$	$\lvert 1\rangle\,\lvert f_i(1)\rangle$
$\lvert 1\rangle\,\lvert 1\rangle$	$\lvert 1\rangle\,\lvert f_i(1)\oplus 1\rangle$

NOTE: $\oplus$ is XOR.

Figure 4.1. Deutsch's algorithm.

Table 4.1. Classical approach.

IN $\lvert x\rangle\lvert y\rangle$	OUT $\lvert x\rangle\lvert f_i(x)\oplus y\rangle$
$\lvert 0\rangle\lvert 0\rangle$	$\lvert 0\rangle\lvert f_0(0)\rangle = \lvert 0\rangle\lvert 0\rangle$
	$\lvert 0\rangle\lvert f_1(0)\rangle = \lvert 0\rangle\lvert 0\rangle$
	$\lvert 0\rangle\lvert f_2(0)\rangle = \lvert 0\rangle\lvert 1\rangle$
	$\lvert 0\rangle\lvert f_3(0)\rangle = \lvert 0\rangle\lvert 1\rangle$
$\lvert 0\rangle\lvert 1\rangle$	$\lvert 0\rangle\lvert f_0(x)\oplus 1\rangle = \lvert 0\rangle\lvert 0\oplus 1\rangle = \lvert 0\rangle\lvert 1\rangle$
	$\lvert 0\rangle\lvert f_1(x)\oplus 1\rangle = \lvert 0\rangle\lvert 0\oplus 1\rangle = \lvert 0\rangle\lvert 1\rangle$
	$\lvert 0\rangle\lvert f_2(x)\oplus 1\rangle = \lvert 0\rangle\lvert 1\oplus 1\rangle = \lvert 0\rangle\lvert 0\rangle$
	$\lvert 0\rangle\lvert f_3(x)\oplus 1\rangle = \lvert 0\rangle\lvert 1\oplus 1\rangle = \lvert 0\rangle\lvert 0\rangle$
$\lvert 1\rangle\lvert 0\rangle$	$\lvert 1\rangle\lvert f_0(1)\rangle = \lvert 1\rangle\lvert 0\rangle$
	$\lvert 1\rangle\lvert f_1(1)\rangle = \lvert 1\rangle\lvert 1\rangle$
	$\lvert 1\rangle\lvert f_2(1)\rangle = \lvert 1\rangle\lvert\rangle 0$
	$\lvert 1\rangle\lvert f_3(1)\rangle = \lvert 1\rangle\lvert 1\rangle$
$\lvert 1\rangle\lvert 1\rangle$	$\lvert 1\rangle\lvert f_0(1)\oplus 1\rangle = \lvert 1\rangle\lvert 0\oplus 1\rangle = \lvert 1\rangle\lvert 0\rangle$
	$\lvert 1\rangle\lvert f_1(1)\oplus 1\rangle = \lvert 1\rangle\lvert 0\oplus 1\rangle = \lvert 1\rangle\lvert 0\rangle$
	$\lvert 1\rangle\lvert f_2(1)\oplus 1\rangle = \lvert 1\rangle\lvert 1\oplus 1\rangle = \lvert 1\rangle\lvert 1\rangle$
	$\lvert 1\rangle\lvert f_3(1)\oplus 1\rangle = \lvert 1\rangle\lvert 1\oplus 1\rangle = \lvert 1\rangle\lvert 0\rangle$

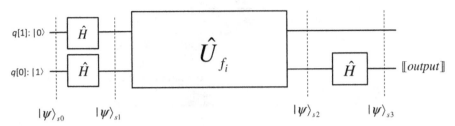

Figure 4.2. Deutsch's gate.

Deutsch showed that, using superposed states of |0> and |1>, we only need to use the gate once, a dramatic reduction of computation. As we pointed out, the H-gate can superpose two qbits (section 3.2.3). Figure 4.2 illustrates the Deutsch gate. Below is the gate analysis. For simplicity, we omit the qbit line numbers.

Step-by-step analysis

$$|\psi\rangle_{s0} = |0\rangle|1\rangle = |01\rangle,$$

$$|\psi\rangle_{s1} = \left[\frac{1}{\sqrt{2}}(|0\rangle + |1\rangle)\right]\left[\frac{1}{\sqrt{2}}(|0\rangle - |1\rangle)\right] = \frac{1}{2}(|00\rangle - |01\rangle + |10\rangle - |11\rangle),$$

$$|\psi\rangle_{s2} = \hat{U}_f|\psi\rangle = s_1\frac{1}{2}(|0\rangle|f_i(0)\rangle - |0\rangle|f_i(0) \oplus 1\rangle + |1\rangle|f_i(1)\rangle - |1\rangle|f_i(1) \oplus 1\rangle)$$

$$= \frac{1}{2}[|0\rangle(|f_i(0)\rangle - |f_i(0) \oplus 1\rangle) + |1\rangle(|f_i(1)\rangle - |f_i(1) \oplus 1\rangle)].$$

It can be shown that

$$|f_i(0)\rangle - |f_i(0) \oplus 1\rangle = (-1)^{f_i(0)}(|0\rangle - |1\rangle) \text{ and } |f_i(1)\rangle - |f_i(1) \oplus 1\rangle$$
$$= (-1)^{f_i(1)}(|0\rangle - |1\rangle). \tag{4.1}$$

Thus,

$$|\psi\rangle_{s2} = \frac{1}{2}\left[(-1)^{f_i(0)}|0\rangle(|0\rangle - |1\rangle) + (-1)^{f_i(1)}|1\rangle(|0\rangle - |1\rangle)\right]$$

$$= \frac{1}{\sqrt{2}}((-1)^{f_i(0)}|0\rangle + (-1)^{f_i(1)}|1\rangle)\frac{1}{\sqrt{2}}(|0\rangle - |1\rangle)$$

$$= \pm \frac{1}{\sqrt{2}}(|0\rangle + |1\rangle)\frac{1}{\sqrt{2}}(|0\rangle - |1\rangle) \quad \text{if } f_i(0) = f_i(1), \text{ and}$$

$$= \pm \frac{1}{\sqrt{2}}(|0\rangle - |1\rangle)\frac{1}{\sqrt{2}}(|0\rangle - |1\rangle) \quad \text{if } f_i(0) \neq f_i(1).$$

Because $\hat{H}^{-1} = \hat{H}$, i.e., $\hat{H}\left(\frac{1}{\sqrt{2}}(|0\rangle + |1\rangle)\right) = |0\rangle$ and $\hat{H}\left(\frac{1}{\sqrt{2}}(|0\rangle - |1\rangle)\right) = |1\rangle$, we obtain

$$|\psi\rangle_{s3} = \pm|0\rangle\frac{1}{\sqrt{2}}(|0\rangle - |1\rangle) \quad \text{if } f_i(0) = f_i(1),$$

$$= \pm|1\rangle\frac{1}{\sqrt{2}}(|0\rangle - |1\rangle) \quad \text{if } f_i(0) \neq f_i(1). \tag{4.2}$$

Therefore the Deutsch's gate outputs the results listed in table 4.2.

Table 4.2. Summary of Deutsch's gate.

i	Observation	Result
0	$|0\rangle \to 0$	Observation of '0' means f_0 or f_3.
1	$|1\rangle \to 1$	
2	$-|1\rangle \to 1$	Observation of '1' means f_1 or f_2.
3	$-|0\rangle \to 1$	

Blueqat codes and outputs

Figure 4.3 shows an implementation of Deutsch's algorithm by Blueqat [6]. The CNOT-gates (section 3.3.1) are used to construct the oracle.

Deutsch's algorithm can be generalized to $N=2^n$ by adding $|0\rangle$ states. For a detailed discussion, refer to Deutsch–Jozsa's algorithm [7].

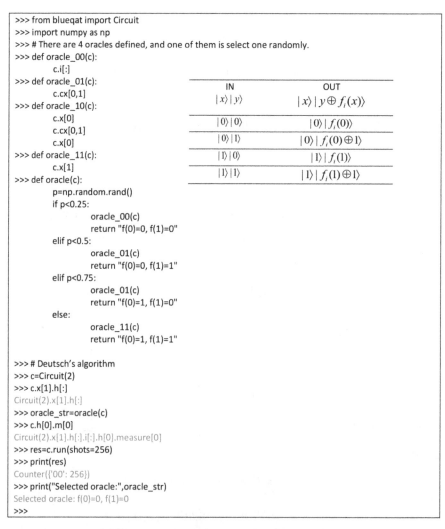

```
>>> from blueqat import Circuit
>>> import numpy as np
>>> # There are 4 oracles defined, and one of them is select one randomly.
>>> def oracle_00(c):
        c.i[:]
>>> def oracle_01(c):
        c.cx[0,1]
>>> def oracle_10(c):
        c.x[0]
        c.cx[0,1]
        c.x[0]
>>> def oracle_11(c):
        c.x[1]
>>> def oracle(c):
        p=np.random.rand()
        if p<0.25:
                oracle_00(c)
                return "f(0)=0, f(1)=0"
        elif p<0.5:
                oracle_01(c)
                return "f(0)=0, f(1)=1"
        elif p<0.75:
                oracle_01(c)
                return "f(0)=1, f(1)=0"
        else:
                oracle_11(c)
                return "f(0)=1, f(1)=1"

>>> # Deutsch's algorithm
>>> c=Circuit(2)
>>> c.x[1].h[:]
Circuit(2).x[1].h[:]
>>> oracle_str=oracle(c)
>>> c.h[0].m[0]
Circuit(2).x[1].h[:].i[:].h[0].measure[0]
>>> res=c.run(shots=256)
>>> print(res)
Counter({'00': 256})
>>> print("Selected oracle:",oracle_str)
Selected oracle: f(0)=0, f(1)=0
>>>
```

IN	OUT
$|x\rangle\,|y\rangle$	$|x\rangle\,|y \oplus f_i(x)\rangle$
$|0\rangle\,|0\rangle$	$|0\rangle\,|f_i(0)\rangle$
$|0\rangle\,|1\rangle$	$|0\rangle\,|f_i(0)\oplus 1\rangle$
$|1\rangle\,|0\rangle$	$|1\rangle\,|f_i(1)\rangle$
$|1\rangle\,|1\rangle$	$|1\rangle\,|f_i(1)\oplus 1\rangle$

Figure 4.3. Blueqat program of Deutsch's algorithm.

4.2 Grover's algorithm

Grover's algorithm is a searching algorithm from an unsorted database [8]. Here is a simple task that can be applied to the algorithm:

Define a binary function $f(xy)$ where $f(00)=0$, $f(01)=0$, $f(10)=0$, and $f(11)=1$ for the binary number array, $xy=00$, 01, 10, and 11. How many times do we need to evaluate $f(xy)$ in order to find out $f(11)=1$?

Classically, it is the same as the expectation value of the number of flipping the four cards, i.e., $(1)(1/4)+(2)(3/4)(1/3)+3(3/4)(2/3) = 2.25$.

Lov Grover applied the amplitude amplification method for efficient searching where the probability amplitude of the target was marked to be negative by applying a 'flipping' operation, and then amplified it by applying a 'amplification' gate. Suppose, for the above binary function, consider the superposed 2-qbits, $|\psi\rangle = \frac{1}{2}(|00\rangle + |01\rangle + |10\rangle + |11\rangle)$ from which we want to search $|11\rangle$. Figure 4.4 illustrates the amplitude and amplification method consisting of the following steps.

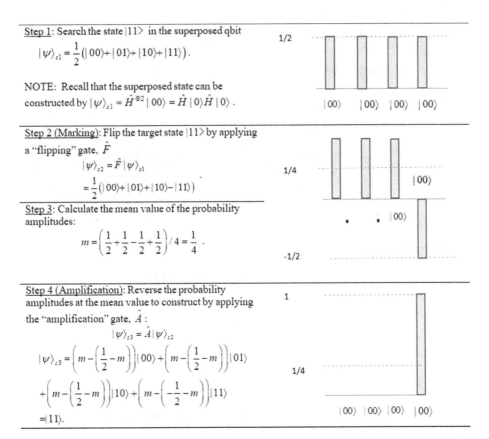

Step 1: Search the state $|11\rangle$ in the superposed qbit
$$|\psi\rangle_{s1} = \frac{1}{2}(|00\rangle + |01\rangle + |10\rangle + |11\rangle).$$

NOTE: Recall that the superposed state can be constructed by $|\psi\rangle_{s1} = \hat{H}^{\otimes 2}|00\rangle = \hat{H}|0\rangle\hat{H}|0\rangle$.

Step 2 (Marking): Flip the target state $|11\rangle$ by applying a "flipping" gate, $\hat{F}$
$$|\psi\rangle_{s2} = \hat{F}|\psi\rangle_{s1}$$
$$= \frac{1}{2}(|00\rangle + |01\rangle + |10\rangle - |11\rangle)$$

Step 3: Calculate the mean value of the probability amplitudes:
$$m = \left(\frac{1}{2} + \frac{1}{2} - \frac{1}{2} + \frac{1}{2}\right)/4 = \frac{1}{4}.$$

Step 4 (Amplification): Reverse the probability amplitudes at the mean value to construct by applying the "amplification" gate, $\hat{A}$:
$$|\psi\rangle_{s3} = \hat{A}|\psi\rangle_{s2}$$
$$|\psi\rangle_{s3} = \left(m - \left(\frac{1}{2} - m\right)\right)|00\rangle + \left(m - \left(\frac{1}{2} - m\right)\right)|01\rangle$$
$$+ \left(m - \left(\frac{1}{2} - m\right)\right)|10\rangle + \left(m - \left(-\frac{1}{2} - m\right)\right)|11\rangle$$
$$= |11\rangle.$$

Figure 4.4. Grover's algorithm.

Marking gate: Steps 1 and 2 of figure 4.4 are given by the following circuit where the controlled-Z (U_{CZ}) gate, where the control bit is $q[1]$ and the target bit is $q[0]$, is used (section 3.3.2).

Marking gate

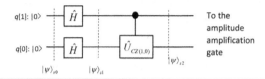

Step-by-step analysis

$$|\psi\rangle_{s0} = |0_1\rangle|0_0\rangle = |0\rangle|0\rangle,$$

$$|\psi\rangle_{s1} = \hat{H}^{\otimes 2}|0\rangle|0\rangle = \left(\frac{|0\rangle + |1\rangle}{\sqrt{2}}\right)\left(\frac{|0\rangle + |1\rangle}{\sqrt{2}}\right)$$

$$= \frac{1}{2}(|00\rangle + |01\rangle + |10\rangle + |11\rangle),$$

$$|\psi\rangle_{s2} = \hat{U}_{CZ(1,0)}|\psi\rangle_{s1} = \frac{1}{2}\hat{U}_{CZ(1,0)}(|00\rangle + |01\rangle + |10\rangle + |11\rangle)$$

$$= \frac{1}{2}(|00\rangle + |01\rangle + |10\rangle - |11\rangle).$$

Amplitude amplification gate: Steps 3 and 4 of figure 4.4 are given by

Amplitude amplification gate

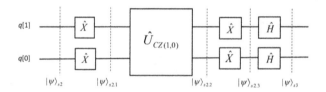

Step-by-step analysis

$$|\psi\rangle_{s2} = \frac{1}{2}(|00\rangle + |01\rangle + |10\rangle - |11\rangle),$$

$$|\psi\rangle_{s2.1} = \hat{X}^{\otimes 2}|\psi\rangle_{s2} = \frac{1}{2}(|11\rangle + |10\rangle + |01\rangle - |00\rangle),$$

$$|\psi\rangle_{s2.2} = \hat{U}_{CZ(1,0)}|\psi\rangle_{s2.1} = \frac{1}{2}(-|11\rangle + |10\rangle + |01\rangle - |00\rangle),$$

$$|\psi\rangle_{s2.3} = \hat{X}^{\otimes 2}|\psi\rangle_{s2.2} = \frac{1}{2}(|00\rangle + |01\rangle + |10\rangle - |11\rangle),$$

$$|\psi\rangle_{s3} = \hat{H}^{\otimes 2}|\psi\rangle_{s2.3}$$

$$= \frac{1}{2}\left[-\left(\frac{|0\rangle+|1\rangle}{\sqrt{2}}\right)\left(\frac{|0\rangle+|1\rangle}{\sqrt{2}}\right) + \left(\frac{|0\rangle+|1\rangle}{\sqrt{2}}\right)\left(\frac{|0\rangle-|1\rangle}{\sqrt{2}}\right)\right.$$

$$\left. + \left(\frac{|0\rangle-|1\rangle}{\sqrt{2}}\right)\left(\frac{|0\rangle+|1\rangle}{\sqrt{2}}\right) - \left(\frac{|0\rangle-|1\rangle}{\sqrt{2}}\right)\left(\frac{|0\rangle-|1\rangle}{\sqrt{2}}\right)\right]$$

$$= -|11\rangle.$$

For marking others, we can apply the following combinations of the S gate and the U_{CZ}-gate shown in figure 4.5.

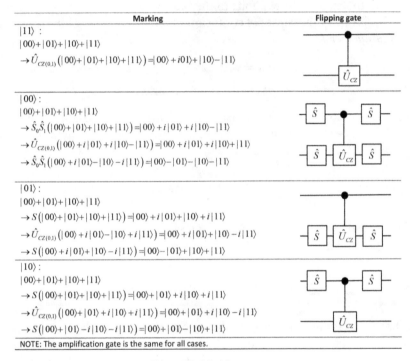

Figure 4.5. Marking gates.

```
>>> from blueqat import Circuit
>>> # Define the common amplitude amplification circuit
>>> a=Circuit(2).h[:].x[:].cz[0,1].x[:].h[:].m[:]
>>> # Check 4 different cases: 00, 01, 10, and 11
>>> #11 Circuit
>>> (Circuit().h[:].cz[0,1]+a).run(shots=200)
Counter({'11': 200})
>>> # 00 Circuit
>>> (Circuit().h[:].s[:].cz[0,1].s[:]+a).run(shots=200)
Counter({'00': 200})
>>> #01 Circuit
>>> (Circuit().h[:].s[1].cz[0,1].s[1]+a).run(shots=200)
Counter({'01': 200})
>>> #10 Circuit
>>> (Circuit().h[:].s[0].cz[0,1].s[0]+a).run(shots=200)
Counter({'10': 200})
>>>
```

Figure 4.6. Grover's algorithm by Blueqat.

Blueqat codes and outputs

In the program listed in figure 4.6, we used an abbreviated code, **+a** which is the amplification circuit:

```
a=Circuit (2).h[:].x[:].cz[0,1].x[:].h[:].m[:]
```

The Blueqat code performs searching as expected.

4.3 Quantum Fourier transform

4.3.1 Idea of quantum Fourier transform (QFT) [9]

In section 4.3 to 4.5, we express qbits in the binary format as well as in the decimal format. (This is one of the causes of confusion.) Here is our *binary number expression of a decimal number k:*

$$k = k_{n-1}k_{n-2}\cdots k_0 = k_{n-1} \cdot 2^{n-1} + k_{n-2} \cdot 2^{n-2} + \cdots$$
$$+ k_0 \cdot 2^0 \text{ where } k_m = 0 \text{ or } 1, \text{ and } N = 2^n \tag{4.3}$$

The discrete *classical* Fourier transform [10] of a set of N data points $x = \{x_j; j = 0, 1, 2, \ldots, N-1\}$ or N-dimensional vector into another N-dimensional vector $y = \{y_j; j = 0, 1, 2, \ldots, N-1\}$ is given by

$$\{x_j\} \rightarrow \{y_j\}: y_j = \frac{1}{\sqrt{N}} \sum_{k=0}^{N-1} x_k e^{i2\pi jk/N} \text{ where we assume the period, } T = 1. \tag{4.4}$$

For the quantum computation of an *n-qbit system*, $\{|k_{n-1}\rangle|k_{n-2}\rangle\cdots|k_0\rangle\}$, where the total number of data points is $N=2^n$, we denote its quantum state vector as

$$|\vec{x}\rangle = \sum_{k_{n-1}=0}^{1} \sum_{k_{n-2}=0}^{1} \cdots \sum_{k_0=0}^{1} x_{k_{n-1}k_{n-2}\cdots k_0}\{|k_{n-1}\rangle|k_{n-2}\rangle\cdots|k_0\rangle\} \text{ or } |\vec{x}\rangle = \sum_{k=0}^{N-1} x_{\{k\}}|k\rangle. \tag{4.5}$$

Notice that the orthonormal basis of the *n*-qbit system is labeled $\{|k\rangle\}=\{|0\rangle, |1\rangle, |2\rangle, \ldots|N-1\rangle\}$ each of which is a qbit in the *decimal* notation. The suffix $\{k\}$ should be addressed using N-numbers: $k_{N-1}\ldots k_1k_0$ (in binary) or k (decimal) is given by equation (4.3). For example, a quantum state vector of a single qbit, where $n=1$ and $N=2$, is given by

$$\sum_{k=0}^{1} x_k|k\rangle = x_0|0\rangle + x_1|1\rangle, \tag{4.6}$$

and a quantum state vector of the 2-qbit system (*n=2* and *N=4*) is

$$\sum_{k_2=0}^{1} \sum_{k_1=0}^{1} x_{k_2k_1}|k_2\rangle|k_1\rangle = x_{00}|0\rangle|0\rangle + x_{01}|0\rangle|1\rangle + x_{10}|1\rangle|0\rangle + x_{11}|1\rangle|1\rangle \text{ (in binary)},$$
$$= x_0|0\rangle + x_1|1\rangle + x_2|2\rangle + x_3|3\rangle \text{ (in decimal)}. \tag{4.7}$$

Now, we define the quantum Fourier transform of quantum state vectors by augmenting the classical Fourier transform given by equation (4.2).

$$|\bar{x}\rangle = \sum_{j=0}^{N-1} x_j |j\rangle \rightarrow |\bar{y}\rangle = \sum_{k=0}^{N-1} y_k |k\rangle$$

$$= \sum_{k=0}^{n-1}\left[\frac{1}{\sqrt{N}}\sum_{j=0}^{N-1}x_k e^{i2\pi jk/N}\right]|k\rangle = \sum_{j=0}^{N-1}x_k\left[\frac{1}{\sqrt{N}}\sum_{k=0}^{N-1}e^{i2\pi jk/N}|k\rangle\right]. \quad (4.8)$$

The above equation can be interpreted that the orthonormal qbit basis $\{|j\rangle\}= \{|j_0\rangle,|j_1\rangle,...,|j_{n-1}\rangle\}$ (decimal) of an n-qbit system is Fourier transformed to another orthonormal qbit basis $\{|k\rangle\}= \{|k_0\rangle,|k_1\rangle,...,|k_{n-1}\rangle\}$ (decimal)

$$|j\rangle \rightarrow \frac{1}{\sqrt{N}}\sum_{k=0}^{N-1}e^{i\frac{2\pi jk}{N}}|k\rangle. \quad (4.9)$$

Examples of QFT
 (1) When $N=2$, the orthonormal basis $\{|j\rangle\}=\{|0\rangle, |1\rangle\}$, each of which are transformed into

$$|0\rangle \rightarrow \frac{1}{\sqrt{2}}\sum_{k=0}^{1}e^{i2\pi\cdot 0\cdot k/2}|k\rangle = \frac{1}{\sqrt{2}}(|0\rangle + |1\rangle), \text{ and } |1\rangle \rightarrow \frac{1}{\sqrt{2}}\sum_{k=0}^{1}e^{i2\pi\cdot 1\cdot k/2}|k\rangle = \frac{1}{\sqrt{2}}(|0\rangle - |1\rangle). \quad (4.10)$$

Notice that this is equivalent to the H-gate operation.
Remark: It can be shown that

$$\hat{H}|j\rangle = \frac{1}{\sqrt{2}}(|0\rangle + e^{i2\pi 0.j}|1\rangle) \quad (4.11)$$

where $|j\rangle = |0\rangle$ or $|1\rangle$, and '$0.j$' appearing in the exponent of equation (4.11) is the 'binary fraction' defined for the *binary number less than 1*:

$$0.j_m j_{m-1}\cdots j_0 = \frac{j_m}{2^1} + \frac{j_{m-2}}{2^2} + \cdots + \frac{j_0}{2^m}. \quad (4.12)$$

 Proof. If $j=0$, $e^{i2\pi 0.j} = e^0 = 1$; and if $j=1$, $e^{i2\pi 0.j} = e^{i2\pi 0.1} = e^{i\pi} = -1$ because 0.1 (binary)=1/2(decimal).
 Thus,
 $|0\rangle + e^{i2\pi 0.j}|1\rangle = |0\rangle + |1\rangle$ if $j=0$; and $|0\rangle + e^{i2\pi 0.j}|1\rangle = |0\rangle - |1\rangle$ if $j=0$.
 (2) When $N=4$, the orthonormal basis of 2-qbit system is given by $\{|0\rangle=|0_1\rangle|0_0\rangle$, $|1\rangle=|0_1\rangle|1_0\rangle, |2\rangle=|1_1\rangle|0_0\rangle, |3\rangle=|1_1\rangle|1_0\rangle\}$, each of which is transformed into:

$$|0\rangle \rightarrow \frac{1}{\sqrt{4}}\sum_{k=0}^{3}e^{i2\pi\cdot 0\cdot k/4}|k\rangle = \frac{1}{2}(|0\rangle + |1\rangle + |2\rangle + |3\rangle),$$

$$|1\rangle \rightarrow \frac{1}{\sqrt{4}}\sum_{k=0}^{3}e^{i2\pi\cdot 1\cdot k/4}|k\rangle = \frac{1}{2}(|0\rangle + i|1\rangle - |2\rangle - i|3\rangle), \quad (4.13)$$

$$|2\rangle \rightarrow \frac{1}{\sqrt{4}}\sum_{k=0}^{3}e^{i2\pi\cdot 2k\cdot/4}|k\rangle = \frac{1}{2}(|0\rangle - |1\rangle + |2\rangle - |3\rangle),$$

$$|3\rangle \rightarrow \frac{1}{\sqrt{4}} \sum_{k=0}^{3} e^{i2\pi \cdot 3 \cdot k/4}|k\rangle = \frac{1}{2}(|0\rangle - i|1\rangle - |2\rangle + i|3\rangle).$$

(3) QFT of $|9\rangle = |1001\rangle$ where $N = 2^4 = 16$ is given by

$$|9\rangle \rightarrow \frac{1}{\sqrt{16}} \sum_{k=0}^{15} e^{i\frac{2\pi 9 k}{16}}|k\rangle$$

$$= \frac{1}{4}(|0\rangle - e^{i\frac{\pi}{8}}|1\rangle + e^{i\frac{\pi}{4}}|2\rangle - e^{i\frac{3\pi}{8}}|3\rangle + e^{i\frac{\pi}{2}}|4\rangle - e^{i\frac{5\pi}{8}}|5\rangle + e^{i\frac{3\pi}{4}}|6\rangle - e^{i\frac{7\pi}{8}}|7\rangle \qquad (4.14)$$

$$- |8\rangle + e^{i\frac{\pi}{8}}|9\rangle - e^{i\frac{\pi}{4}}|10\rangle + e^{i\frac{3\pi}{8}}|11\rangle - e^{i\frac{\pi}{2}}|12\rangle + e^{i\frac{5\pi}{8}}|13\rangle - e^{i\frac{3\pi}{4}}|14\rangle + e^{i\frac{7\pi}{8}}|15\rangle)$$

where $e^{i9\pi/8} = e^{i\pi + i\pi/8} = -e^{i\pi/8}$ etc.

4.3.2 QFT of orthogonal basis

In order to apply QFT in a quantum computer, *the qbit basis needs to be Fourier transformed* given by equation (4.9). For the *n*-qbit system ($N = 2^n$), QFT is given by

$$|j\rangle \xrightarrow[\text{QF}]{} \frac{1}{\sqrt{2^n}} \sum_{k=0}^{2^n-1} e^{i\frac{2\pi jk}{2^n}}|k\rangle = \frac{1}{\sqrt{2^n}}\left(|0\rangle + e^{i2\pi j\frac{1}{2}}|1\rangle\right)_0 \left(|0\rangle + e^{i2\pi j\frac{1}{2^2}}|1\rangle\right)_1 \cdots$$

$$\left(|0\rangle + e^{i2\pi j\frac{1}{2^n}}|1\rangle\right)_{n-1}. \qquad (4.15)$$

Proof. From equations (4.3), (4.5), (4.9), and using the binary format, we obtain

$$|j\rangle \xrightarrow[\text{QF}]{} \frac{1}{\sqrt{2^n}} \sum_{k=0}^{2^n-1} e^{i\frac{2\pi jk}{2^n}}|k\rangle = \frac{1}{\sqrt{2^n}} \sum_{k_{n-1}=0}^{1} \sum_{k_{n-2}=0}^{1} \cdots$$

$$\sum_{k_0=0}^{1} e^{i\frac{2\pi j(k_{n-1}2^{n-1}+k_{n-2}2^{n-2}+\cdots+k_0 2^0)}{2^n}}|k_{n-1}\rangle|k_{n-2}\rangle\cdots|k_0\rangle$$

$$= \frac{1}{\sqrt{2^n}} \sum_{k_{n-1}=0}^{1} \sum_{k_{n-2}=0}^{1} \cdots \sum_{k_0=0}^{1} e^{i2\pi j\left(\frac{k_{n-1}}{2}+\frac{k_{n-2}}{2^2}+\cdots+\frac{k_0}{2^n}\right)}|k_{n-1}\rangle|k_{n-2}\rangle\cdots|k_0\rangle$$

$$= \frac{1}{\sqrt{2^n}} \sum_{k_{n-1}=0}^{1} \sum_{k_{n-2}=0}^{1} \cdots \sum_{k_0=0}^{1} \left(e^{i2\pi j\frac{k_1}{2}}|k_{n-1}\rangle\right)_{n-1}\left(e^{i2\pi j\frac{k_2}{2^2}}|k_{n-2}\rangle\right)_{n-2}\cdots\left(e^{i2\pi j\frac{k_n}{2^n}}|k_0\rangle\right)_0 \qquad (4.16)$$

$$= \frac{1}{\sqrt{2^n}}\left(\sum_{k_{n-1}=0}^{1} e^{i2\pi j\frac{k_{n-1}}{2}}|k_1\rangle\right)_{n-1}\left(\sum_{k_{n-2}=0}^{1} e^{i2\pi j\frac{k_{n-2}}{2^2}}|k_{n-2}\rangle\right)_{n-2}\cdots\left(\sum_{k_n=0}^{1} e^{i2\pi j\frac{k_0}{2^n}}|k_0\rangle\right)_0$$

$$= \frac{1}{\sqrt{2^n}}\left(|0\rangle + e^{i2\pi j\frac{1}{2}}|1\rangle\right)_{n-1}\left(|0\rangle + e^{i2\pi j\frac{1}{2^2}}|1\rangle\right)_{n-2}\cdots\left(|0\rangle + e^{i2\pi j\frac{1}{2^n}}|1\rangle\right)_0.$$

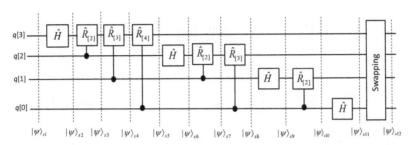

Figure 4.7. Quantum Fourier transform ($n=4$).

Figure 4.7 illustrates the QFT circuit for $N=4$ where the gate $R_{[m]}$ is the controlled phase rotation gate (section 3.3.2) defined as

$$\hat{R}_{[m]}|0\rangle = |0\rangle \text{ and } \hat{R}_{[m]}|1\rangle = e^{i2\pi/2^m}|1\rangle. \tag{4.17}$$

Step-by-step analysis

Below is the analysis of QFT of $|9\rangle = |1001\rangle$. Because the qbit line numbers are important, we keep them in the analysis below.

$$|\psi\rangle_{s1} = |9\rangle = |1_3\rangle|0_2\rangle|0_1\rangle|\rangle 1_0,$$

$$|\psi\rangle_{s2} = \hat{H}_3|\psi\rangle_{s1} = \frac{1}{\sqrt{2}}(|0\rangle - |1\rangle)_3|0_2\rangle|0_1\rangle|1_0\rangle,$$

$$|\psi\rangle_{s3} = \hat{R}_{[2]}|\psi\rangle_{s2} = \frac{1}{\sqrt{2}}(|0\rangle - |1\rangle)_3|0_2\rangle|0_1\rangle|1_0\rangle,$$

$$|\psi\rangle_{s4} = \hat{R}_{[3]}|\psi\rangle_{s3} = \frac{1}{\sqrt{2}}(|0\rangle - |1\rangle)_3|0_2\rangle|0_1\rangle|1_0\rangle,$$

$$|\psi\rangle_{s5} = \hat{R}_{[4]}|\psi\rangle_{s4} = \frac{1}{\sqrt{2}}\left(|0_3\rangle - e^{i\frac{\pi}{8}}|1_3\rangle\right)|0_2\rangle|0_1\rangle|1_0\rangle,$$

$$|\psi\rangle_{s6} = \hat{H}_2|\psi\rangle_{s5} = \frac{1}{\sqrt{2}}\left(|0_3\rangle - e^{i\frac{\pi}{8}}|1_3\rangle\right)\frac{1}{\sqrt{2}}(|0\rangle + |1\rangle)_2|\rangle 0_1|1_0\rangle,$$

$$|\psi\rangle_{s7} = \hat{R}_{[2]}|\psi\rangle_{s6} = \frac{1}{\sqrt{2}}\left(|0_3\rangle - e^{i\frac{\pi}{8}}|1_3\rangle\right)\frac{1}{\sqrt{2}}(|0\rangle + |1\rangle)_2|0_1\rangle|1_0\rangle,$$

$$|\psi\rangle_{s8} = \hat{R}_{[3]}|\psi\rangle_{s7} = \frac{1}{\sqrt{2}}\left(|0\rangle - e^{i\frac{\pi}{8}}|1\rangle\right)_3\frac{1}{\sqrt{2}}\left(|0\rangle + e^{i\frac{\pi}{4}}|1\rangle\right)_2|0_1\rangle|1\rangle_0,$$

$$|\psi\rangle_{s9} = \hat{H}_1|\psi\rangle_{s8} = \frac{1}{\sqrt{2}}\left(|0\rangle - e^{i\frac{\pi}{8}}|1\rangle\right)_3\frac{1}{\sqrt{2}}\left(|0\rangle + e^{i\frac{\pi}{4}}|1\rangle\right)_2\frac{1}{\sqrt{2}}(|0\rangle + |1\rangle)_2|1_0\rangle,$$

$$|\psi\rangle_{s10} = \hat{R}_{[2]}|\psi\rangle_{s9} = \frac{1}{\sqrt{2}}\left(|0\rangle - e^{i\frac{\pi}{8}}|1\rangle\right)_3\frac{1}{\sqrt{2}}\left(|0\rangle + e^{i\frac{\pi}{4}}|1\rangle\right)_2\frac{1}{\sqrt{2}}\left(|0\rangle + e^{i\frac{\pi}{2}}|1\rangle\right)_1|1_0\rangle,$$

$$|\psi\rangle_{s11} = \hat{H}_0|\psi\rangle_{s10} = \frac{1}{\sqrt{2}}\left(|0\rangle - e^{i\frac{\pi}{8}}|1\rangle\right)_3 \frac{1}{\sqrt{2}}\left(|0\rangle + e^{i\frac{\pi}{4}}|1\rangle\right)_2 \frac{1}{\sqrt{2}}\left(|0\rangle + e^{i\frac{\pi}{2}}|1\rangle\right)_1 \frac{1}{\sqrt{2}}(|0\rangle - |1\rangle)_0,$$

$$|\psi\rangle_{s12} = [\text{Swapping}]|\psi\rangle_{s11} = \frac{1}{\sqrt{2}}(|0\rangle - |1\rangle)_3 \frac{1}{\sqrt{2}}\left(|0\rangle + e^{i\frac{\pi}{2}}|1\rangle\right)_2 \frac{1}{\sqrt{2}}\left(|0\rangle + e^{i\frac{\pi}{4}}|1\rangle\right)_1 \frac{1}{\sqrt{2}}\left(|0\rangle - e^{i\frac{\pi}{8}}|1\rangle\right)_0$$

$$= \frac{1}{4}(|0000\rangle - e^{i\frac{\pi}{8}}|0001\rangle + e^{i\frac{\pi}{4}}|0010\rangle - e^{i\left(\frac{\pi}{4}+\frac{\pi}{8}\right)}|0011\rangle + e^{i\frac{\pi}{2}}|0100\rangle - e^{i\left(\frac{\pi}{2}+\frac{\pi}{8}\right)}|0101\rangle$$

$$+ e^{i\left(\frac{\pi}{2}+\frac{\pi}{4}\right)}|0110\rangle - e^{i\left(\frac{\pi}{2}+\frac{\pi}{4}+\frac{\pi}{8}\right)}|0111\rangle - |1000\rangle + e^{i\frac{\pi}{8}}|1001\rangle - e^{i\frac{\pi}{4}}|1010\rangle$$

$$+ e^{i\left(\frac{\pi}{4}+\frac{\pi}{8}\right)}|1011\rangle - e^{i\frac{\pi}{2}}|1100\rangle + e^{i\left(\frac{\pi}{2}+\frac{\pi}{8}\right)}|1101\rangle - e^{i\left(\frac{\pi}{2}+\frac{\pi}{4}\right)}|1110\rangle + e^{i\left(\frac{\pi}{2}+\frac{\pi}{4}+\frac{\pi}{8}\right)}|1111\rangle)$$

$$= \frac{1}{4}(|0\rangle - e^{i\frac{\pi}{8}}|1\rangle + e^{i\frac{\pi}{4}}|2\rangle - e^{i\frac{3\pi}{8}}|3\rangle + e^{i\frac{\pi}{2}}|4\rangle - e^{i\frac{5\pi}{8}}|5\rangle + e^{i\frac{3\pi}{4}}|6\rangle - e^{i\frac{7\pi}{8}}|7\rangle$$

$$- |8\rangle + e^{i\frac{\pi}{8}}|9\rangle - e^{i\frac{\pi}{4}}|10\rangle + e^{i\frac{3\pi}{8}}|11\rangle - e^{i\frac{\pi}{2}}|12\rangle + e^{i\frac{5\pi}{8}}|13\rangle - e^{i\frac{3\pi}{4}}|14\rangle + e^{i\frac{7\pi}{8}}|15\rangle).$$

This is the same result as equation (4.14), and the circuit is correctly assembled.

Blueqat codes and outputs

The controlled phase rotation of Blueqat is given by `.cphase(math.pi/m)` where m = integer (section 3.3.2). For the swapping gate, we use the swap gate `.SWAP[a,b]` (section 3.3.4).

Notice that the output from the Blueqat code is binary and of the order of |000>, |0001>, ..., |1111>.

The n-qbit QFT circuit is given by the quantum circuit depicted by figure 4.9.

Step-by-step analysis

$|\psi\rangle_{s0} = |j_{n-1}\rangle|j_{n-2}\rangle\cdots|j_0\rangle$ where $|j_k\rangle = |0\rangle$ or $|1\rangle$, $k=n-1, n-2, ...,1, 0$.

We use the expression of the H-gate operation given by equation (4.11)

$$|\psi\rangle_{s1} = \hat{H}_{n-1}|j_{n-1}\rangle|j_{n-2}\rangle\cdots|j_0\rangle = \frac{1}{\sqrt{2}}(|0\rangle + e^{i2\pi 0.j_{n-1}}|1\rangle)_{n-1}|j_2\rangle\cdots n - 2\cdots|j\rangle_0.$$

```
>>> #QFT of |9>=|1001>
>>> from blueqat import Circuit
>>> import math
>>>
Circuit().x[:].h[3].cphase(math.pi/2)[2,3].cphase(math.pi/4)[1,3].cphase(math.pi/8)[0,3].h[2].cphase(math.
pi/2)[1,2].cphase(math.pi/4)[0,2].h[1].cphase(math.pi/2)[0,1].h[0].swap[0,3].swap[1,2].run()
array([ 2.50000000e-01+0.j ,      2.30969883e-01-0.09567086j ,
       1.76776695e-01-0.1767767j ,  9.56708581e-02-0.23096988j ,
      -1.53080850e-17-0.25j ,      - 9.56708581e-02-0.23096988j ,
      -1.76776695e-01-0.1767767j , -2.30969883e-01-0.09567086j ,
      -2.50000000e-01+0.j ,       -2.30969883e-01+0.09567086j ,
      -1.76776695e-01+0.1767767j , -9.56708581e-02+0.23096988j ,
       1.53080850e-17+0.25j ,       9.56708581e-02+0.23096988j ,
       1.76776695e-01+0.1767767j ,  2.30969883e-01+0.09567086j])
>>>
```

Figure 4.8. QFT of |9>=|1001>.

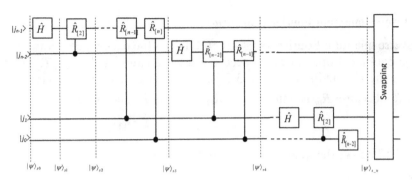

Figure 4.9. Quantum Fourier transform circuit ($N=2^n$) .

Apply the controlled phase rotation gate $R_{[2]}$ to obtain

$$|\psi\rangle_{s2} = \hat{R}_{[2]}|\psi\rangle_{s1} = \frac{1}{\sqrt{2}}(|0\rangle + e^{i2\pi 0.j_{n-1}j_{n-2}}|1\rangle)_{n-1}|j_{n-2}\rangle|j_{n-3}\rangle\cdots\cdots|j_0\rangle$$

where $0.j_{n-1}j_{n-2}$ is a binary fraction (equation (4.12)). This is not so obvious and we should prove it.

Proof. If $j_{n-2}=0$, $e^{i2\pi 0.j_{n-1}j_{n-2}} = e^{i2\pi 0.j_{n-1}}$; and if $j_{n-2}=1$, $e^{i2\pi 0.j_{n-1}}e^{i\frac{2\pi}{2^2}} = e^{i2\pi(0.j_{n-1}+\frac{1}{2^2})}$ $= e^{i2\pi 0.j_{n-1}j_{n-2}}$.
Thus,

$$|\psi\rangle_{s2} = \hat{R}_{[2]}|\psi\rangle_{s1} = \frac{1}{\sqrt{2}}\left(|0\rangle + e^{i2\pi 0.j_{n-1}j_{n-2}}|1\rangle\right)_{n-1}|j_{n-2}\rangle|j_{n-3}\rangle\cdots\cdots|j_0\rangle.$$

Repeat the operators $R_{[2]}$ to $R_{[n]}$ on $|j_{n-1}\rangle$, and we obtain

$$|\psi\rangle_{s3} = \hat{R}_{[n]}\hat{R}_{[n-1]}\cdots\hat{R}_{[2]}|j\rangle_{s1} = \frac{1}{\sqrt{2}}\left(|0\rangle + e^{i2\pi 0.j_{n-1}j_{n-2}\cdots j_0}|1\rangle\right)_{n-1}|j_{n-2}\rangle|j_{n-3}\rangle\cdots\cdots|j_0\rangle.$$

Operators on $|j_{n-2}\rangle$ are similar, and we obtain

$$|\psi\rangle_{s4} = \frac{1}{\sqrt{2}}\left(|0\rangle + e^{i2\pi 0.j_{n-1}j_{n-2}\cdots j_0}|1\rangle\right)_{n-1}\frac{1}{\sqrt{2}}\left(|0\rangle + e^{i2\pi 0.j_{n-2}j_{n-3}\cdots j_0}|1\rangle\right)_{n-2}|j_{n-3}\rangle\cdots\cdots|j_0\rangle,$$

and then

$$|\psi\rangle_{sn} = \frac{1}{\sqrt{2^n}}\left(|0\rangle + e^{i2\pi 0.j_{n-1}j_{n-2}\cdots j_0}|1\rangle\right)_{n-1}\left(|0\rangle + e^{i2\pi 0.j_{n-2}j_{n-3}\cdots j_0}|1\rangle\right)_{n-2}\cdots\cdots\left(|0\rangle + e^{i2\pi 0.j_0}|1\rangle\right)_0.$$

After the swapping gates, QFT is completed.

$$|j\rangle \rightarrow \frac{1}{\sqrt{2^n}}\left(|0\rangle + e^{i2\pi 0.j_0}|1\rangle\right)_{n-1}\left(|0\rangle + e^{i2\pi 0.j_1 j_0}|1\rangle\right)_{n-2}\cdots\cdots\left(|0\rangle + e^{i2\pi 0.j_{n-1}j_{n-2}\cdots j_0}|1\rangle\right)_0$$

$$= \frac{1}{\sqrt{2^n}}\left(|0\rangle + e^{i2\pi\frac{j_0}{2}}|1\rangle\right)_{n-1}\left(|0\rangle + e^{i2\pi\left(\frac{j_1}{2}+\frac{j_0}{2^2}\right)}|1\rangle\right)_{n-2}\cdots\cdots\left(|0\rangle + e^{i2\pi\left(\frac{j_{n-1}}{2}+\frac{j_{n-2}}{2^2}+\cdots+\frac{j_0}{2^{n-1}}\right)}|1\rangle\right)_0. \quad (4.18)$$

4.3.3 Inverse quantum Fourier transform

The inverse quantum Fourier transform (QFT^{-1}) is the backward operation of QFT. For the backward operation, we need the reverse gates used in QFT. The reverse H-gate is the H-gate itself (section 3.2.2). The reversed controlled phase rotation is $\hat{R}_m^{-1} = \hat{R}_m^+$ because $\hat{R}_m$ is unitary, i.e., $\hat{R}_{[m]} \cdot \hat{R}_{[m]}^+ = \hat{I}$. Therefore,

$$\hat{R}_{[m]}^+|0\rangle = |0\rangle, \text{ and } \hat{R}_{[m]}^+|1\rangle = e^{-i2\pi/2^m}|1\rangle. \tag{4.19}$$

Figure 4.10 shows the quantum circuit of QFT^{-1} of $N=4$.

Step-by-step analysis

$QFT^{-1}|2\rangle = QFT^{-1}|0_3\rangle|0_2\rangle|1_1\rangle|0_0\rangle$ is perfomed as follows.

$$|\psi\rangle_{s0} = |2\rangle = |0_3\rangle|0_2\rangle|1_1\rangle|0_0\rangle,$$

$$|\psi\rangle_{s1} = [\mathrm{Swapping}]|\psi\rangle_{s0} = |0_3\rangle|1_2\rangle|0_1\rangle|0_0\rangle,$$

$$|\psi\rangle_{s2} = \hat{H}_0|\psi\rangle_{s1} = |0_3\rangle|1_2\rangle|0_1\rangle\hat{H}_0|0_0\rangle = |0_3\rangle|1_2\rangle|0_1\rangle\frac{1}{\sqrt{2}}(|0\rangle + |1\rangle)_0,$$

$$|\psi\rangle_{s3} = \hat{R}_{[2]}^+|\psi\rangle_{s2} = |0_3\rangle|1_2\rangle\hat{R}_{[2]}^+|0_1\rangle\frac{1}{\sqrt{2}}(|0\rangle + |1\rangle)_0$$

$$= \frac{1}{\sqrt{2}}|0_3\rangle|1_2\rangle\hat{R}_{[2]}^+|0_1\rangle(|0\rangle + |1\rangle)_0 = \frac{1}{\sqrt{2}}|0\rangle|1\rangle|0\rangle(|0\rangle + |1\rangle)_0,$$

$$|\psi\rangle_{s4} = \hat{H}_1|\psi\rangle_{s3} = |0_3\rangle|1_2\rangle(\hat{H}_1|0_1\rangle)\frac{1}{\sqrt{2}}(|0\rangle + |1\rangle)_0$$

$$= |0_3\rangle|1_2\rangle\frac{1}{\sqrt{2}}(|0\rangle + |1\rangle)_1\frac{1}{\sqrt{2}}(|0\rangle + |1\rangle)_0,$$

$$|\psi\rangle_{s5} = \hat{R}_{[3]}^+|\psi\rangle_{s4} = |0_3\rangle\hat{R}_{[3]}^+|1_2\rangle\frac{1}{\sqrt{2}}(|0\rangle + |1\rangle)_1\frac{1}{\sqrt{2}}(|0\rangle + |1\rangle)_0$$

$$= \frac{1}{2}[|0_3\rangle|1_2\rangle(|0\rangle + |1\rangle)_1|0_0\rangle + |0\rangle_3 e^{-i\frac{2\pi}{2^3}}|1_2\rangle(|0\rangle + |1\rangle)_1|1\rangle_0]$$

$$= \frac{1}{2}[|0_3\rangle|\rangle 1_2(|0\rangle + |1\rangle)_1|0_0\rangle + |0_3\rangle|1_2\rangle(|0\rangle + |1\rangle)_2(e^{-i\frac{2\pi}{2^3}}|1\rangle)_0]$$

$$= \frac{1}{2}\left[|0_0\rangle|1_2\rangle(|0\rangle + |1\rangle)_1\left(|0\rangle + e^{-i\frac{2\pi}{2^3}}|1\rangle\right)_0\right],$$

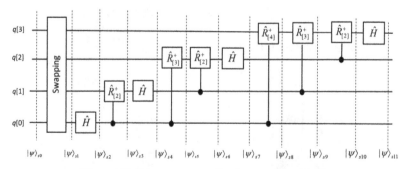

Figure 4.10. Inverse quantum Fourier transform ($N=4$).

```
>>> from blueqat import Circuit
>>> import math
>>> Circuit().x[1].swap[0,3].swap[1,2].h[0].cphase(-math.pi/2)[0,1].h[1].cphase(-math.pi/4)[0,2].cphase(-
math.pi/2)[1,2].h[2].cphase(-math.pi/8)[0,3].cphase(-math.pi/4)[1,3].cphase(-math.pi/2)[2,3].h[3].run()
array([ 2.50000000e-01+0.j    , 1.76776695e-01-0.1767767j ,
        1.53080850e-17-0.25j   , -1.76776695e-01-0.1767767j ,
       -2.50000000e-01+0.j    , -1.76776695e-01+0.1767767j ,
       -1.53080850e-17+0.25j   , 1.76776695e-01+0.1767767j ,
        2.50000000e-01+0.j    , 1.76776695e-01-0.1767767j ,
        1.53080850e-17-0.25j   , -1.76776695e-01-0.1767767j ,
       -2.50000000e-01+0.j    , -1.76776695e-01-0.1767767j,
       -1.53080850e-17+0.25j   , 1.76776695e-01+0.1767767j])
>>>
```

Figure 4.11. Blueqat's inverse QFT ($N=4$).

$$|\psi\rangle_{s7} = \hat{H}|\psi\rangle_{s6} = \frac{1}{2}|0_3\rangle\frac{1}{\sqrt{2}}(|0\rangle - |1\rangle)_2\left(|0\rangle + e^{-i\frac{2\pi}{2^2}}|1\rangle\right)_1\left(|0\rangle + e^{-i\frac{2\pi}{2^3}}|1\rangle\right)_0,$$

$$|\psi\rangle_{s8} = \hat{R}_{[4]}^+|\psi\rangle_{s7} = \frac{1}{2}|0_3\rangle\frac{1}{\sqrt{2}}(|0\rangle - |1\rangle)_2\left(|0\rangle + e^{-i\frac{2\pi}{2^2}}|1\rangle\right)_1\left(|0\rangle + e^{-i\frac{2\pi}{2^3}}|1\rangle\right)_0,$$

$$|\psi\rangle_{s9} = \hat{R}_{[3]}^+|\psi\rangle_{s8} = \frac{1}{2}|0_3\rangle\frac{1}{\sqrt{2}}(|0\rangle - |1\rangle)_2\left(|0\rangle + e^{-i\frac{2\pi}{2^2}}|1\rangle\right)_1\left(|0\rangle + e^{-i\frac{2\pi}{2^3}}|1\rangle\right)_0,$$

$$|\psi\rangle_{s10} = \hat{R}_{[2]}^+|\psi\rangle_{s9} = \frac{1}{2}|0_0\rangle\frac{1}{\sqrt{2}}(|0\rangle - |1\rangle)_2\left(|0\rangle + e^{-i\frac{2\pi}{2^2}}|1\rangle\right)_1\left(|0\rangle + e^{-i\frac{2\pi}{2^3}}|1\rangle\right)_0,$$

$$|\psi\rangle_{s11} = \hat{H}|\psi\rangle_{s10} = \frac{1}{4}\left(|0\rangle + |1\rangle\right)_3\left(|0\rangle - |1\rangle\right)_2(|0\rangle + e^{-i\frac{2\pi}{2^2}}|1\rangle)_1\left(|0\rangle + e^{-i\frac{2\pi}{2^3}}|1\rangle\right)_0$$

$$= \frac{1}{4}\left(|000\rangle + e^{-i\frac{\pi}{4}}|0001\rangle - i|0010\rangle - ie^{-i\frac{\pi}{4}}|0011\rangle - |0100\rangle - e^{-i\frac{\pi}{4}}|0101\rangle + i|0110\rangle\right.$$

$$+ ie^{-i\frac{\pi}{4}}|0111\rangle$$

$$+ |1000\rangle + e^{-i\frac{\pi}{4}}|1001\rangle - i|1010\rangle - ie^{-i\frac{\pi}{4}}|1011\rangle - |1100\rangle - e^{-i\frac{\pi}{4}}|1101\rangle + ie^{-i\frac{\pi}{4}}|1111\rangle).$$

Blueqat codes and outputs

Figure 4.11 shows Blueqat's $QFT^{-1}|2\rangle = QFT^{-1}|0010\rangle$. The first code .x[1] is used to generate $|0010\rangle$. The reversed controlled phase rotations are the CP-gate itself with negative angles (section 3.3.2). The output is in the order of $|0000\rangle$, $|0001\rangle$, ..., $|1111\rangle$. Note $-1.53080850e-17$ is essentially zero.

4.4 Phase estimation

The phase estimation is to find phase ϕ ($0 \leqslant \phi < 1$) such that $\hat{U}|u\rangle = e^{i2\pi\phi}|u\rangle$ where $\hat{U}$ is a *controlled* unitary operator, and $|u\rangle$ is its eigen function. Shor's algorithm will be discussed in the next section that needs this circuit. In the figure 4.12, $\hat{U}^{2^k} = \hat{U}\hat{U}\cdots\hat{U}$ ($k = 0, 1, 2, ..., n-1$) and $\hat{U}^{2^k}|u\rangle = e^{i2\pi 2^k\phi}|u\rangle$ when the control bit is $|1\rangle$. In figure 4.12, the symbol, [obs], denotes measurement of quantum state.

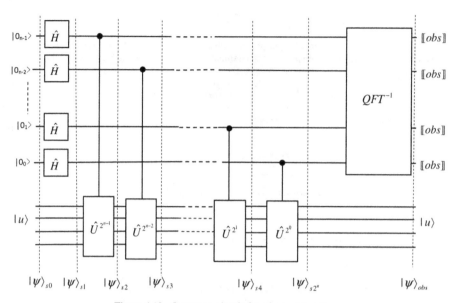

Figure 4.12. Quantum circuit for phase estimation.

Step-by-step analysis

$$|\psi\rangle_{s0} = |0_{n-1}\rangle|0_{n-2}\rangle\cdots\cdots|0_0\rangle|u\rangle,$$

$$|\psi\rangle_{s1} = \left(\hat{H}|0\rangle\right)\left(\hat{H}|0\rangle\right)\cdots\cdots\left(\hat{H}|0\rangle\right)|u\rangle$$

$$= \frac{1}{\sqrt{2^n}}(|0\rangle + |1\rangle)(|0\rangle + |1\rangle)\cdots\cdots(|0\rangle + |1\rangle)|u\rangle,$$

$$|\psi\rangle_{s2} = \frac{1}{\sqrt{2^n}}\left(|0\rangle + |1\rangle\right)\left(|0\rangle + |1\rangle\right)\cdots\cdots\left(|0\rangle + |1\rangle\right)\hat{U}^{2^{n-1}}|u\rangle$$

$$= \frac{1}{\sqrt{2^n}}\left(|0\rangle + e^{i2\pi 2^{n-1}\phi}|1\rangle\right)\left(|0\rangle + |1\rangle\right)\cdots\cdots\left(|0\rangle + |1\rangle\right)|u\rangle,$$

$$|\psi\rangle_{s3} = \frac{1}{\sqrt{2^n}}\left(|0\rangle + |1\rangle\right)\left(|0\rangle + |1\rangle\right)\cdots\cdots\left(|0\rangle + |1\rangle\right)\hat{U}^{2^2}|u\rangle$$

$$= \frac{1}{\sqrt{2^n}}\left(|0\rangle + e^{i2\pi 2^{n-1}\phi}|1\rangle\right)\left(|0\rangle + e^{i2\pi 2^{n-2}\phi}|1\rangle\right)\cdots\cdots\left(|0\rangle + |1\rangle\right)\left(|0\rangle + |1\rangle\right)|u\rangle,$$

$$|\psi\rangle_{s4} = \frac{1}{\sqrt{2^n}}\left(|0\rangle + e^{i2\pi 2^{n-1}\phi}|1\rangle\right)\left(|0\rangle + e^{i2\pi 2^{n-2}\phi}|1\rangle\right)\cdots\cdots\left(|0\rangle + |1\rangle\right)\left(|0\rangle + |1\rangle\right)\hat{U}^{2^1}|u\rangle$$

$$= \frac{1}{\sqrt{2^n}}\left(|0\rangle + e^{i2\pi 2^{n-1}\phi}|1\rangle\right)\left(|0\rangle + e^{i2\pi 2^{n-2}\phi}|1\rangle\right)\cdots\cdots\left(|0\rangle + e^{i2\pi 2^1\phi}|1\rangle\right)\left(|0\rangle + |1\rangle\right)|u\rangle,$$

$$|\psi\rangle_{s2^n} = \frac{1}{\sqrt{2^n}}\left(|0\rangle + e^{i2\pi 2^{n-1}\phi}|1\rangle\right)\left(|0\rangle + e^{i2\pi 2^{n-2}\phi}|1\rangle\right)\cdots\cdots\left(|0\rangle + e^{i2\pi 2^1\phi}|1\rangle\right)\left(|0\rangle + |1\rangle\right)\hat{U}^{2^0}|u\rangle$$

$$= \frac{1}{\sqrt{2^n}}\left(|0\rangle + e^{i2\pi 2^{n-1}\phi}|1\rangle\right)\left(|0\rangle + e^{i2\pi 2^{n-2}\phi}|1\rangle\right)\cdots\cdots\left(|0\rangle + e^{i2\pi 2^0\phi}|1\rangle\right)|u\rangle$$

$$= \frac{1}{\sqrt{2^n}}\left[|0\rangle + e^{i2\pi 2^n\left(\frac{\phi}{2}\right)}|1\rangle\right]\left[|0\rangle + e^{i2\pi 2^n\left(\frac{\phi}{2^2}\right)}|1\rangle\right]\cdots\cdots\left(|0\rangle + e^{i2\pi 2^n\left(\frac{\phi}{2^n}\right)}|1\rangle\right)|u\rangle$$

$$= \frac{1}{\sqrt{2^n}}\left[|0\rangle + e^{i2\pi\left(\frac{\tilde{\phi}}{2}\right)}|1\rangle\right]\left[|0\rangle + e^{i2\pi\left(\frac{\tilde{\phi}}{2^2}\right)}|1\rangle\right]\cdots\cdots\left(|0\rangle + e^{i2\pi\left(\frac{\tilde{\phi}}{2^n}\right)}|1\rangle\right)|u\rangle.$$

where $\tilde{\phi} = 2^n\phi$. Comparing the *QFT* formula of equation (4.18), $|\psi\rangle_{s2^n}$ has the *QFT* of $\tilde{\phi}$, and thus

$$|\psi\rangle_{sN} = \left(\frac{1}{\sqrt{2^n}}\sum_{k=0}^{2^n-1} e^{i2\pi\tilde{\phi}k/N}|k\rangle\right)|u\rangle \tag{4.20}$$

in the decimal notation where $N=2^n$.

Therefore, the last stage of the phase estimation circuit, QFT^{-1}, finds

$$|\tilde{\phi}\rangle = 2^n|\phi\rangle. \tag{4.21}$$

4.5 Shor's algorithm for prime factorization

The most sophisticated encryption technology at present, Rivest–Shamir–Adleman (RSA) encryption, is a public-key cryptosystem that utilizes the multiplication of two large prime numbers (2048 bits) because prime factoring is an astronomically time-consuming task even using a super computer. However, Shor pointed out that prime factoring can be performed easily using quantum computation [11, 12].

The basic idea of prime factoring is that a number has periodicity and if we find the period of a number, factoring the number can be completed. Shor's algorithm applies the inverse quantum Fourier transform (QFT^{-1}) in order to find the periodicity of a number. In this section, we explain what the periodicity of a number is and how QFT^{-1} can be used to find the periodicity.

4.5.1 Periodicity of a number

Let us define $F_n = x^n \pmod{M}$ where n is an integer and M is a product of two prime numbers, $M=p\cdot q$. F_n needs to calculate the remainder after x^n is divided by M. Here is the procedure of prime factoring:

1) $F_n = x^n \pmod{M}$ always has a periodic property. Let us calculate $x \pmod{M}$, $x^2 \pmod{M}$, $x^3 \pmod{M}$, ... Then we may find r such that $x \pmod{M} = x^r \pmod{M}$ $=1$. This r can be called the 'rank.' For example, consider a number $M=15$ which is a product of prime numbers 3 and 5, and $F_n = 2^n \pmod{15}$ or 'F_n is the reminder after 2^n is divided by 15 where $n = 0, 1, 2, 3, 4, 5, 6, 7, ...$' Table 4.3 lists the rank of number 15. From this table, we find the 'period' of the number 15 is 4: $r = 4$.

Table 4.3. Rank of number 15.

F_n	2^n (mod 15)	Reminder
F_0	1 (mod 15)	1
F_1	2 (mod 15)	2
F_2	4 (mod 15)	4
F_3	8 (mod 15)	8
F_4	16 (mod 15)	1
F_5	32 (mod 15)	2
F_6	64 (mod 15)	4
F_7	128 (mod 15)	8

2) If $F_n = x^n$ (mod M)=1, then there is an integer r that satisfies $x^n = rM+1$. Thus, $x^n - 1 = rM$, which means M has a common divisor r. Therefore, once you find the period r, the number M can be factored! For example, in the case of the number 15, because $r=4$ and $x=2$, we obtain $x^r - 1 = 2^4 - 1 = (2^2+1)(2^2-1) = (5)(3)$! However, if $M = p \cdot q$ is an extremely large number of 300 digits or so, prime factoring of finding find p and q is a formidable task.

4.5.2 Shor's idea

Shor's algorithm finds r by calculating x^n (mod M) all at once using quantum entanglement. First, we explain the general formula of Shor's algorithm, then demonstrate it by factoring 15. IBM did it using their quantum computer in 2001.

In order to find the rank, r, consider an operator, $\hat{U}$, such that such that

$$\hat{U}|y\rangle \equiv |xy(\text{mod } M)\rangle \tag{4.22}$$

and its eigen function, $|u\rangle$, and eigen value, τ, i.e., $\hat{U}|u\rangle = \tau|u\rangle$. We now show that the eigen value is $\tau = e^{i2\pi\frac{s}{r}}$, and the eigen function is given by

$$|u_s\rangle = \frac{1}{\sqrt{r}} \sum_{j=0}^{r-1} e^{-i2\pi j\frac{s}{r}}|x^j(\text{mod } M)\rangle. \tag{4.23}$$

Thus, $\hat{U}|u\rangle = e^{i2\pi\frac{s}{r}}|u\rangle$, $s = 0, 1, 2, ..., r$.

There are several pre-steps required to prove the above equations.

1) Proof of $\hat{U}|x^j(\text{mod } M)\rangle = |x^{j+1}(\text{mod } M)\rangle$.

Let $y = x^j(\text{mod } M)$ in $\hat{U}|y\rangle = |xy(\text{mod } M)\rangle$, i.e., $\hat{U}|x^j(\text{mod } M)\rangle = |x(x^j(\text{mod } M)(\text{mod } M))\rangle$.

Suppose $x^j = pM + q$ where p and q are natural numbers. Because $xpM(\text{mod } M) = 0$,

$$\hat{U}|x^j(\text{mod } M)\rangle = |x(pM + q)(\text{mod } M)\rangle = |xq(\text{mod } M)\rangle.$$

On the other hand,

$$|x^{j+1}(\bmod M)\rangle = |x \cdot x^j(\bmod M)\rangle$$
$$= |x(pM + q)(\bmod M)\rangle = |xpM(\bmod M)\rangle + |xq(\bmod M)\rangle$$
$$= |xq(\bmod M)\rangle$$

Therefore, $\hat{U}|x^j(\bmod M)\rangle = |x^{j+1}(\bmod M)\rangle$.

2) Proof of $\hat{U}|u\rangle = e^{i2\pi\frac{s}{r}}|u\rangle$.

If $|u_s\rangle = \frac{1}{\sqrt{r}}\sum_{j=0}^{r-1}e^{-i2\pi j\frac{s}{r}}|x^j(\bmod M)\rangle$, then $\hat{U}|u_s\rangle = \frac{1}{\sqrt{r}}\sum_{j=0}^{r-1}e^{-i2\pi j\frac{s}{r}}\hat{U}|x^j(\bmod M)\rangle$.

Because $\hat{U}|x^j(\bmod M)\rangle = |x^{j+1}(\bmod M)\rangle$ as we proved in step (1),

$$\hat{U}|u_s\rangle = \frac{1}{\sqrt{r}}\sum_{j=0}^{r-1}e^{-i2\pi j\frac{s}{r}}\hat{U}|x^j(\bmod M)\rangle = \frac{1}{\sqrt{r}}\sum_{j=0}^{r-1}e^{-i2\pi j\frac{s}{r}}|x^{j+1}(\bmod M)\rangle$$

$$= \frac{1}{\sqrt{r}}[e^{-i2\pi\cdot 0\cdot\frac{s}{r}}|x^1(\bmod M)\rangle + e^{-i2\pi\cdot 1\cdot\frac{s}{r}}|x^2(\bmod M)\rangle + \cdots$$

$$+ e^{-i2\pi\cdot(r-1)\cdot\frac{s}{r}}|x^r(\bmod M)\rangle]$$

$$= e^{i2\pi\frac{s}{r}}\frac{1}{\sqrt{r}}[e^{-i2\pi\cdot 1\cdot\frac{s}{r}}|x^1(\bmod M)\rangle + e^{-i2\pi\cdot 2\cdot\frac{s}{r}}|x^2(\bmod M)\rangle + \cdots$$

$$+ e^{-i2\pi\cdot r\cdot\frac{s}{r}}|x^r(\bmod M)\rangle].$$

Now, because $x^r(\bmod M) = 1 = x^0$ and $e^{-i2\pi\cdot r\frac{s}{r}} = 1 = e^{-i2\pi\cdot 0\cdot\frac{s}{r}}$, the last term appearing in the above can be changed to

$$e^{-i2\pi\cdot r\cdot\frac{s}{r}}|x^r(\bmod M)\rangle = e^{-i2\pi\cdot 0\cdot\frac{s}{r}}|x^0(\bmod M)\rangle. \tag{4.24}$$

Therefore, by moving the last term to the front, we obtain

$$\hat{U}|u_s\rangle = e^{i2\pi\frac{s}{r}}\frac{1}{\sqrt{r}}[e^{-i2\pi\cdot 1\cdot\frac{s}{r}}|x^1(\bmod M)\rangle + e^{-i2\pi\cdot 2\cdot\frac{s}{r}}|x^2(\bmod M)\rangle + \cdots$$

$$\cdots\cdots + e^{-i2\pi\cdot(r-1)\cdot\frac{s}{r}}|x^r(\bmod M)\rangle + e^{-i2\pi\cdot 0\cdot\frac{s}{r}}|x^0(\bmod M)\rangle]$$

$$= e^{i2\pi\frac{s}{r}}\frac{1}{\sqrt{r}}[e^{-i2\pi\cdot 0\cdot\frac{s}{r}}|x^0(\bmod M)\rangle + e^{-i2\pi\cdot 1\cdot\frac{s}{r}}|x^1(\bmod M)\rangle + e^{-i2\pi\cdot 2\cdot\frac{s}{r}}|x^2(\bmod M)\rangle+ \tag{4.25}$$

$$\cdots\cdots + e^{-i2\pi\cdot(r-1)\cdot\frac{s}{r}}|x^r(\bmod M)\rangle]$$

$$= e^{i2\pi\frac{s}{r}}|u\rangle.$$

3) If we solve $\hat{U}|u_s\rangle = e^{i2\pi\frac{s}{r}}|u\rangle$, we can get r for prime factoring! Recall that we have the phase estimation circuit to find phase ϕ $(0 \leqslant \phi < 1)$ such that $\hat{U}|u\rangle = e^{i2\pi\phi}|u\rangle$ (figure 4.12). However, notice that, because $|u\rangle$ is unknown, we cannot operate the circuit unless we know r. Nevertheless, the sum of all the eigen functions always satisfies $\frac{1}{\sqrt{r}}\sum_{s=0}^{r-1}|u_s\rangle = |1\rangle$ as we prove in the next step, and we use |1> instead of using individual $|u\rangle$.

4) Proof of $\dfrac{1}{\sqrt{r}}\displaystyle\sum_{s=0}^{r-1}|u_s\rangle = |1\rangle$.

$$\frac{1}{\sqrt{r}}\sum_{s=0}^{r-1}|u_s\rangle = \frac{1}{\sqrt{r}}\sum_{s=0}^{r-1}\frac{1}{\sqrt{r}}\sum_{j=0}^{r-1}e^{-i2\pi j\frac{s}{r}}|x^j(\mathrm{mod}\,M)\rangle = \frac{1}{r}\sum_{j=0}^{r-1}\left(\sum_{s=0}^{r-1}e^{-i2\pi j\frac{s}{r}}\right)|x^j(\mathrm{mod}\,M)\rangle$$

$$= \frac{1}{r}\sum_{j=0}^{r-1}(r\delta_{j,\,0})|x^j(\mathrm{mod}\,M)\rangle = \frac{1}{r}r|x^0(\mathrm{mod}\,M)\rangle = |1\rangle$$

where $\delta_{0,\,j}$ is Kronecker's delta.

5) Using equation (4.23) and the above step (4), after operating all control gates of the phase estimation circuit (figure 4.12), $\hat{U}^{2^0}\cdots\hat{U}^{2^{n-1}}$, we obtain

$$\left(\frac{1}{\sqrt{2^n}}\sum_{k=0}^{2^n-1}|k\rangle\right)\hat{U}^k|1\rangle = \frac{1}{\sqrt{N}}\sum_{k=0}^{2^n-1}|k\rangle|x^k(\mathrm{mod}\,M)\rangle. \qquad (4.26)$$

Notice that this means that the controlled gates calculate all $\{x^k(\mathrm{mod}\,M), k=0, 1, 2, \ldots, N\}$ all at once where $2^n=N$.

Using $\dfrac{1}{\sqrt{r}}\displaystyle\sum_{s=0}^{r-1}|u_s\rangle = |1\rangle$, the above equation (4.26) becomes

$$\frac{1}{\sqrt{N}}\sum_{k=0}^{2^n-1}|k\rangle|x^k(\mathrm{mod}\,M)\rangle = \left(\frac{1}{\sqrt{2^n}}\sum_{k=0}^{2^n-1}|k\rangle\right)\hat{U}^k|1\rangle$$

$$= \frac{1}{\sqrt{2^n}}\sum_{k=0}^{2^n-1}|k\rangle\left(\hat{U}^k\frac{1}{\sqrt{r}}\sum_{s=0}^{r-1}|u_s\rangle\right) = \frac{1}{\sqrt{r}}\sum_{s=0}^{r-1}\frac{1}{\sqrt{2^n}}\sum_{k=0}^{2^n-1}|k\rangle\left(\hat{U}^k|u_s\rangle\right)$$

$$= \frac{1}{\sqrt{r}}\sum_{s=0}^{r-1}\frac{1}{\sqrt{2^n}}\sum_{k=0}^{2^n-1}|k\rangle\left(e^{i2\pi k\frac{s}{r}}|u_s\rangle\right) = \frac{1}{\sqrt{r}}\sum_{s=0}^{r-1}\left(\frac{1}{\sqrt{2^n}}\sum_{k=0}^{2^n-1}e^{i2\pi k\frac{s}{r}}|k\rangle\right)|u_s\rangle$$

$$= \frac{1}{\sqrt{r}}\sum_{s=0}^{r-1}\left(\frac{1}{\sqrt{2^n}}\sum_{k=0}^{2^n-1}e^{i2\pi k\frac{\tilde{s}}{r}\cdot\frac{1}{2^n}}|k\rangle\right)|u_s\rangle = \frac{1}{\sqrt{r}}\sum_{s=0}^{r-1}\left(\frac{1}{\sqrt{N}}\sum_{k=0}^{N-1}e^{i2\pi k\frac{\tilde{s}}{r}\cdot\frac{1}{N}}|k\rangle\right)|u_s\rangle.$$

Notice that $\dfrac{1}{\sqrt{N}}\displaystyle\sum_{k=0}^{N-1}e^{i2\pi k\frac{\tilde{s}}{r}\cdot\frac{1}{N}}|k\rangle$ is the QFT of the ratio $\dfrac{\tilde{s}}{r}$ by referring to equation (4.9). In other words,

$$\frac{\tilde{s}}{r} = QFT^{-1}\left[\frac{1}{\sqrt{N}}\sum_{k=0}^{N-1}e^{i2\pi k\frac{\tilde{s}}{r}\cdot\frac{1}{N}}|k\rangle\right]$$

Therefore, once we obtain $s/r = (\tilde{s}/rN)$, we can estimate r!

4.5.3 Prime factorization of $M=15$

How can we perform the prime factorization of $M=15$? Use figure 4.12 where $n=4$ and $N=16$, after operating the controlled gates, and we obtain

$$|\psi\rangle = \left(\frac{1}{\sqrt{2^n}} \sum_{k=0}^{N-1} |k\rangle\right) \hat{U}^k |1\rangle = \frac{1}{\sqrt{2^n}} \sum_{k=0}^{2^n-1} |k\rangle |2^k (\mathrm{mod}\ 15)\rangle \qquad (4.27)$$

where we selected $x=2$.

For n, it must be $2^n \geqslant M$. We can take $n=4$ for $M=15$, and we calculate all possible $|x^k (\mathrm{mod}\ M)\rangle$:

$$|\psi\rangle = \frac{1}{4} \sum_{k=0}^{2^n-1} |k\rangle |2^k (\mathrm{mod}\ 15)\rangle = \frac{1}{4} (|0\rangle |2^0 (\mathrm{mod}\ 15)\rangle + |1\rangle |2^1 (\mathrm{mod}\ 15)\rangle + \cdots +$$

$$|0\rangle |2^{15} (\mathrm{mod}\ 15)\rangle)$$

$$= \frac{1}{4} (|0\rangle |1\rangle + |1\rangle |2\rangle + |3\rangle |8\rangle + |4\rangle |1\rangle + |5\rangle |2\rangle + |6\rangle |4\rangle + |7\rangle |8\rangle \qquad (4.28)$$

$$+ |8\rangle |1\rangle + |9\rangle |2\rangle + |10\rangle |4\rangle + |11\rangle |8\rangle + |12\rangle |1\rangle + |13\rangle |2\rangle + |14\rangle |4\rangle + |15\rangle |8\rangle)$$

$$= \frac{1}{4} [(|0\rangle + |4\rangle + |8\rangle + |12\rangle) |1\rangle + (|1\rangle + |5\rangle + |9\rangle + |13\rangle) |2\rangle$$

$$+ (|2\rangle + |6\rangle + |10\rangle + |14\rangle) |4\rangle + (|3\rangle + |7\rangle + |11\rangle + |15\rangle) |8\rangle].$$

By performing QFT^{-1}, we obtain

$$\begin{aligned}
|0\rangle + |4\rangle + |8\rangle + |12\rangle &\to |0\rangle + |4\rangle + |8\rangle + |12\rangle, \\
|1\rangle + |5\rangle + |9\rangle + |13\rangle &\to |0\rangle - i|4\rangle - |8\rangle + i|12\rangle, \\
|2\rangle + |6\rangle + |10\rangle + |14\rangle &\to |0\rangle - |4\rangle + |8\rangle - |12\rangle, \\
3\rangle + |7\rangle + |11\rangle + |15\rangle &\to |0\rangle + i|4\rangle - |8\rangle - i|12\rangle.
\end{aligned} \qquad (4.29)$$

Figure 4.19 at the end of this chapter lists QFT^{-1} of each term of the above equations. Thus,

$$|\psi\rangle = \frac{1}{4} \sum_{k=0}^{2^n-1} |k\rangle |2^k (\mathrm{mod}\ 15)\rangle = \frac{1}{4} (|0\rangle |2^0 (\mathrm{mod}\ 15)\rangle + |1\rangle |2^1 (\mathrm{mod}\ 15)\rangle$$

$$+ \cdots + |0\rangle |2^{15} (\mathrm{mod}\ 15)\rangle)$$

$$= \frac{1}{4} [(|0\rangle + |4\rangle + |8\rangle + |12\rangle) |1\rangle + (|1\rangle + |5\rangle + |9\rangle + |13\rangle)) \qquad (4.30)$$

$$|2\rangle + (|2\rangle + |6\rangle + |10\rangle + |14\rangle) |4\rangle + (|3\rangle + |7\rangle + |11\rangle + |15\rangle) |8\rangle]$$

$$= (|0\rangle + |4\rangle + |8\rangle + |12\rangle) |1\rangle + (|0\rangle - i|4\rangle - |8\rangle + i|12\rangle) |2\rangle$$

$$+ (|0\rangle - |4\rangle + |8\rangle - |12\rangle) |4\rangle + (|0\rangle + i|4\rangle - |8\rangle - i|12\rangle) |8\rangle.$$

From this result, possible $\tilde{s}/r$ are 0, 4, 8, or 12. To obtain the rank r, these numbers need to be divided by $2^4=16$, and $s/r = 0$, (1/4), (1/2), (3/4). Thus, the possible r will be 2 or 4 because odd numbers are excluded. Recall $x^r (\mathrm{mod}\ M)=1$, and because 2^2 (mod 15) $\neq 1$, $r=2$ is impossible. Therefore, the rank of $M=15$ is $r=4$. Using $r=4$ and $x=2$, we obtain $x^r-1=2^4-1=(2^2+1)(2^2-1)=(5)(3)$!

4.6 Simon's algorithm

Simon's algorithm provides the first example of exponential speedup over the classical algorithm by using a quantum computation to solve a particular problem [13]. Suppose there is a mathematical function that maps binary strings of length n to another binary string of the same length such that $f(x) = f(y)$ if and only if $y = x$ (one-to-one) or $y = x \oplus s$ (two-to-one) where s is a secret binary string of length n.
The goal for Simon's algorithm is to :
Effectively determine if f is one-to-one or two-to-one by finding the secret string s.
For example, suppose the secret key is $s = 110$, we obtain

$$000 \oplus 110=110 \quad 001 \oplus 110=111 \quad 010 \oplus 110=100 \quad 011 \oplus 110=101$$

$$100 \oplus 110=010 \quad 101 \oplus 110=011 \quad 110 \oplus 110=000 \quad 111 \oplus 110=011$$

For this value of $s=110$, $f(000)=f(110)$, $f(001)=f(111)$, $f(010)=f(100)$, and $f(011)=f(101)$.
Thus, the question will be

How may function evaluations need to be determined for the secret string?

If we find, for example, $f(011)=f(101)$, we have $011 \oplus s_2s_1s_0 = 101$ where $s = s_2s_1s_0$. Then the secret string can be determined by
$011 \oplus (011 \oplus s_2s_1s_0) = 011 \oplus 101$, and thus $s_2s_1s_0 = 011 \oplus 101=110$ because $011\oplus011=000$.
The quantum circuit of Simon's algorithm is shown in figure 4.13.
The first Hadamard gates create a superposed qbits:

$$|\psi\rangle_{s0} = |0_{2n-1}\rangle \cdots |0_n\rangle|0_{n-1}\rangle \cdots |0_0\rangle,$$

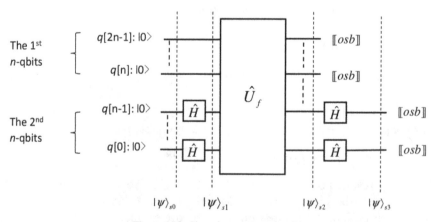

Figure 4.13. Simon's quantum circuit.

$$|\psi\rangle_{s1} = |0_{2n-1}\rangle\cdots|0_n\rangle H^{\otimes n}|0_{n-1}\rangle\cdots|0_0\rangle.$$

The oracle, U_f, calculates $|x\rangle|0\rangle \rightarrow |x\rangle|0 \oplus f(x)\rangle = |x\rangle|f(x)\rangle$, and after the oracle, we obtain

$$|\psi\rangle_{s2} = \frac{1}{\sqrt{2^n}}|f(x)\rangle|x\rangle.$$

Next, measure the first n-qbits. The observed value, $f(x)$ could correspond to two possible inputs, i.e., x and $y=x\oplus s$, and $f(y)=f(x)$. Then, using equation (3.4), the second n-qbits will follow:

(i)

$$\hat{H}^{\otimes n}|x\rangle = \frac{1}{\sqrt{2^n}}\sum_z (-1)^{x\cdot z}|z\rangle \qquad (4.31)$$

if $f(x)$ is one-to-one, or

(ii) $\qquad \hat{H}^{\otimes n}|x\rangle = \frac{1}{\sqrt{2^n}}\sum_z [(-1)^{x\cdot z} + (-1)^{y\cdot z}]|z\rangle$

if $f(x)$ is two-to-one because the second n-qbits will be $\frac{|x\rangle + |y\rangle}{\sqrt{2}}$.

Observation of the second n-qbits will give an output only if $(-1)^{x\cdot z}=(-1)^{y\cdot z}$, otherwise the amplitude, $(-1)^{x\cdot z}+(-1)^{y\cdot z}$, cancels out. This means $x\cdot z=y\cdot z=(x\oplus s)\cdot z$, and therefore $s\cdot z=0$ (mod 2). Repeating the observation enough to find n-independent equations $\{s\cdot z_k=0, k=0, 1, \ldots n\}$. Then we can solve these equations to determine $s=s_{n-1}s_{n-2}\ldots s_0$. Figure 4.14 shows the quantum circuit for $n=3$ and $s=110$. The oracle can be written using the CNOT gate (section 3.3.1).

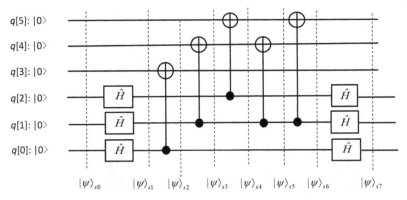

Figure 4.14. Simon's gate ($n=3$ and $s=110$) .

Step-by-step analysis
Following the rule of 'the most significant bit on the top,'

$$|\psi\rangle_{s0} = |0_5\rangle|0_4\rangle|0_3\rangle|0_2\rangle|0_1\rangle|0_0\rangle = |000\rangle|000\rangle,$$

$$|\psi\rangle_{s1} = |000\rangle\frac{1}{\sqrt{2^3}}(|0\rangle + |1\rangle) \otimes (|0\rangle + |1\rangle) \otimes (|0\rangle + |1\rangle)$$

$$=\frac{1}{\sqrt{2^3}}[|000000\rangle + |000010\rangle + |000100\rangle + |000110\rangle + |000001\rangle + |000011\rangle + |000101\rangle + |000111\rangle].$$

$$|\psi\rangle_{s2} = \hat{U}_{CN(0,3)}|\psi\rangle_{s1}$$

$$=\frac{1}{\sqrt{2^3}}[|000000\rangle + |000010\rangle + |000100\rangle + |000110\rangle + |001001\rangle + |001011\rangle + |001101\rangle + |001111\rangle],$$

$$|\psi\rangle_{s3} = \hat{U}_{CN(1,4)}|\psi\rangle_{s2}$$

$$=\frac{1}{\sqrt{2^3}}[|000000\rangle + |010010\rangle + |000100\rangle + |010110\rangle + |001001\rangle + |011011\rangle + |001101\rangle + |011111\rangle],$$

$$|\psi\rangle_{s5} = \hat{U}_{CN(1,4)}|\psi\rangle_{s4}$$

$$=\frac{1}{\sqrt{2^3}}[|000000\rangle + |000010\rangle + |100100\rangle + |100110\rangle + |001001\rangle + |001011\rangle + |101101\rangle + |101111\rangle],$$

$$|\psi\rangle_{s6} = \hat{U}_{CN(1,5)}|\psi\rangle_{s5}$$

$$=\frac{1}{\sqrt{2^3}}[|f(000)\rangle|000\rangle + |f(010)\rangle|010\rangle + |f(100)\rangle|100\rangle + |f(110)\rangle|110\rangle$$

$$+|f(001)\rangle|001\rangle + |f(011)\rangle|011\rangle + |f(101)\rangle|101\rangle + |f(111)\rangle|111\rangle]$$

$$=\frac{1}{\sqrt{2^3}}[|000000\rangle + |100010\rangle + |100100\rangle + |000110\rangle$$

$$+|001001\rangle + |101011\rangle + |101101\rangle + |001111\rangle]$$

$$=\frac{1}{\sqrt{2^3}}[|000\rangle|000\rangle + |100\rangle|010\rangle + |100\rangle|100\rangle + |000\rangle|110\rangle$$

$$+|001\rangle|001\rangle + |101\rangle|011\rangle + |101\rangle|101\rangle + |001\rangle|111\rangle]$$

$$=\frac{1}{\sqrt{2^3}}[|000\rangle(|0\rangle|0\rangle|0\rangle + |1\rangle|1\rangle|0\rangle) + |001\rangle(|0\rangle|0\rangle|1\rangle + |1\rangle|1\rangle|1\rangle)$$

$$+|100\rangle(|0\rangle|1\rangle|0\rangle + |1\rangle|0\rangle|0\rangle) + |101\rangle(|0\rangle|1\rangle|1\rangle + |1\rangle|0\rangle|1\rangle)].$$

We can check what we have so far. From the above equation, we can obtain the secret key $s=s_2s_1s_0$. Table 4.4 lists the secret key for each case of 000, 001, 010, and 011.

Table 4.4. Checking secret key.

From Oracle	$f(x)=f(y)$	Binary string relations: $y=x$ or $y = x \oplus s$
$f(000)=000$ $f(101)=100$	$f(000)=f(110)$	$000 \oplus s_2s_1s_0 = 110$ $s = s_2s_1s_0 = 000 \oplus 110=110$
$f(100)=100$ $f(110)=000$	$f(001)=f(111)$	$001 \oplus s_2s_1s_0 = 111$ $s = s_2s_1s_0 = 001 \oplus 111=110$
$f(001)=001$ $f(011)=100$	$f(010)=f(100)$	$010 \oplus s_2s_1s_0 = 100$ $s = s_2s_1s_0 = 010 \oplus 100=110$
$f(101)=101$ $f(111)=001$	$f(011)=f(101)$	$011 \oplus s_2s_1s_0 = 101$ $s = s_2s_1s_0 = 001 \oplus 101=110$

The secret string s can be determined as follows.

$$|\psi\rangle_{s7} = \widehat{H}^{\otimes 3}|\psi\rangle_{s6}$$

$$= \frac{1}{8}[|000\rangle(|000\rangle + |001\rangle + |110\rangle + |111\rangle) + |001\rangle(|000\rangle - |001\rangle + |100\rangle - |111\rangle)$$

$$+ |100\rangle(|000\rangle + |001\rangle - |110\rangle - |111\rangle) + |101\rangle(|000\rangle - |001\rangle - |110\rangle + |111\rangle)].$$

Therefore, we obtain $s=110$ in following manner:

$s\cdot 000 = s_2\cdot 0 \oplus s_1\cdot 0 \oplus s_0\cdot 0 = 0$	$s=000$ to be excluded
$s\cdot 001 = s_2\cdot 0 \oplus s_1\cdot 0 \oplus s_0\cdot 1 = 0$	$s_0=0$
$s\cdot 110 = s_2\cdot 1 \oplus s_1\cdot 1 = 0$	$s_2\cdot 1 \oplus s_1\cdot 1 = 0$, thus $s_2=s_1=1$. Then $s=110$
$s\cdot 111 = s_2\cdot 1 \oplus s_1\cdot 1 = 0$	$s_2\cdot 1 \oplus s_1\cdot 1 = 0$, the same as above.

Blueqat codes and outputs

We show two cases: $n=2$, $s=11$ and $n=3$, $s=110$ in figure 4.15.

1. For $n=2$, $s=11$, the output is

$$\frac{1}{2}[|00\rangle(|00\rangle + |11\rangle) + |11\rangle(|00\rangle - |11\rangle)] = \frac{1}{2}[|0000\rangle + |0011\rangle + |1100\rangle - |1111\rangle].$$

2. For $n=3$, $s=110$,

$$\frac{1}{4}[|000\rangle(|000\rangle + |001\rangle + |110\rangle + |111\rangle) + |001\rangle(|000\rangle - |001\rangle + |100\rangle - |111\rangle)$$

$$+ |100\rangle(|000\rangle + |001\rangle - |110\rangle - |111\rangle) + |101\rangle(|000\rangle - |001\rangle - |110\rangle + |111\rangle)]$$

$$= \frac{1}{4}[|0000\rangle + |000001\rangle + |000110\rangle + |000111\rangle + |001000\rangle - |001001\rangle + |001100\rangle - |00111\rangle$$

$$+ |100000\rangle + |100001\rangle - |100110\rangle - |100111\rangle + |101000\rangle - |101001\rangle - |101110\rangle + |101111\rangle].$$

```
>>> from blueqat import Circuit
# Simon's algorithm
# n=2; |q(3)q(2)>=|y1y0> and |q(1)q(0)>=|x1x0> where |y>=|x>(+)|s>, |s>=11
>>> Circuit().h[0].h[1].cnot[0,2].cnot[0,3].cnot[1,2].cnot[1,3].h[1].h[0].run()
array([ 0.5+0.j,    0. +0.j,    0. +0.j,    0.5+0.j,    0. +0.j,    0. +0.j,
        0. +0.j,    0. +0.j,    0. +0.j,    0. +0.j,    0. +0.j,    0. +0.j,
        0.5+0.j,    0. +0.j,    0. +0.j,   -0.5+0.j])

>>> # n=3; |q(5)q(4)q(3)>=|y2y1y0> and |q(2)q(1)q(0)>=|x2x1x0> where |y>=|x>(+)|s>, |s>=110
>>> Circuit().h[0].h[1].h[2].cnot[0,3].cnot[1,4].cnot[2,5].cnot[1,4].cnot[1,5].h[2].h[1].h[0].run()
array([ 0.25+0.j,   0.25+0.j,   0. +0.j,    0. +0.j,    0. +0.j,    0. +0.j,
        0.25+0.j,   0.25+0.j,   0.25+0.j,  -0.25+0.j,   0. +0.j,    0. +0.j,
        0. +0.j,    0. +0.j,    0.25+0.j,  -0.25+0.j,   0. +0.j,    0. +0.j,
        0. +0.j,    0. +0.j,    0. +0.j,    0. +0.j,    0. +0.j,    0. +0.j,
        0. +0.j,    0. +0.j,    0. +0.j,    0. +0.j,    0. +0.j,    0. +0.j,
        0. +0.j,    0. +0.j,    0.25+0.j,   0.25+0.j,   0. +0.j,    0. +0.j,
        0. +0.j,    0. +0.j,   -0.25+0.j,  -0.25+0.j,   0.25+0.j,  -0.25+0.j,
        0. +0.j,    0. +0.j,    0. +0.j,    0. +0.j,   -0.25+0.j,   0.25+0.j,
        0. +0.j,    0. +0.j,    0. +0.j,    0. +0.j,    0. +0.j,    0. +0.j,
        0. +0.j,    0. +0.j,    0. +0.j,    0. +0.j,    0. +0.j,    0. +0.j,
        0. +0.j,    0. +0.j,    0. +0.j,    0. +0.j])
>>>
```

Figure 4.15. Simon' algorithm (key: $|s\rangle=110$).

4.7 Error corrections

Whether it is classical or quantum, data error correction is a major issue for accurate and reproducible results. For example, thermal noise would excite or flip spin states very easily. The basic idea of the error correction is based on the law of large numbers, i.e., the correct one must have the highest probability by repeating many trials. However, in quantum physics, measurements collapse the quantum states. Any attempts to measure the state of a quantum system to check for errors could possibly change the state, and correcting errors would be impossible. Nevertheless, there are ideas of quantum error corrections. In this chapter, we only describe bit flip error and phase flip error.

4.7.1 Bit flip error [14]

The bit flip error can be expressed as $a|0\rangle + b|1\rangle \rightarrow a|1\rangle + b|0\rangle$. Suppose Alice wants to send the qbit to Bob, and Bob wants to detect a bit flip and correct it. First, extra bits, which are called *ancilla*, are introduced to add redundancy. For this purpose, an encoded circuit applied to the qbit to be protected to create $|\psi\rangle = a|000\rangle + b|111\rangle$ as shown in figure 4.16.

Step-by-step analysis

There are three possibilities of a bit flip:

(i) $a|0\rangle_0|0\rangle_1|0\rangle_2 + b|1\rangle_0|1\rangle_1|1\rangle_2 \rightarrow a|1\rangle_0|0\rangle_1|0\rangle_2 + b|0\rangle_0|1\rangle_1|1\rangle_2$,

(ii) $a|0\rangle_0|0\rangle_1|0\rangle_2 + b|1\rangle_0|1\rangle_1|1\rangle_2 \rightarrow a|0\rangle_0|1\rangle_1|0\rangle_2 + b|1\rangle_0|0\rangle_1|1\rangle_2$, and

(iii) $a|0\rangle_0|0\rangle_1|0\rangle_2 + b|1\rangle_0|1\rangle_1|1\rangle_2 \rightarrow a|0\rangle_0|0\rangle_1|1\rangle_2 + b|1\rangle_0|1\rangle_1|0\rangle_2$.

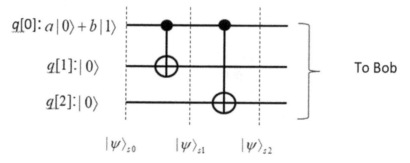

Figure 4.16. Adding redundancy to a qbit.

Table 4.5 shows the three steps shown in the above figure.

Table 4.5. Step-by-step analysis after adding ancilla.

$|\psi\rangle_{s0}$	$|\psi\rangle_{s1} = \hat{U}_{CN(0,1)}|\psi\rangle_{s0}$	$|\psi\rangle_{s2} = \hat{U}_{CN(0,2)}|\psi\rangle_{s1}$
$(a|0\rangle_0 + b|1\rangle_0)|0\rangle_1|0\rangle_2$	$a|0\rangle_0|0\rangle_1|0\rangle_2 + b|1\rangle_0|1\rangle_1|0\rangle_2$	$a|0\rangle_0|0\rangle_1|0\rangle_2 + b|1\rangle_0|1\rangle_1|1\rangle_2$

The third one corresponds to the extra bit flip. Error detection is similar to the classical parity check [11]. For this purpose, we use the parity test circuit illustrated in figure 4.17. The circuit compares the first bit with the second bit, and the first bit with the third bit to detect a bit flip.

Corrective actions

Table 4.6 summarized the corrective action for each possible case.

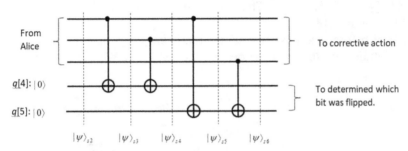

Figure 4.17. Bit flip error correction.

Table 4.6. Bit flip detection.

	No error: $a\|000\rangle + b\|111\rangle$	1st bit flip: $a\|100\rangle + b\|011\rangle$	2nd bit flip: $a\|010\rangle + b\|101\rangle$	3rd bit flip: $a\|001\rangle + b\|110\rangle$
$\|\psi\rangle_{s2}\|0\rangle_3\|0\rangle_4$	$a\|000\rangle_{012}\|0\rangle_3\|0\rangle_4$ $+ b\|111\rangle_{012}\|0\rangle_3\|0\rangle_4$	$a\|100\rangle_{012}\|0\rangle_3\|0\rangle_4$ $+ b\|011\rangle_{012}\|0\rangle_3\|0\rangle_4$	$a\|010\rangle_{012}\|0\rangle_3\|0\rangle_4$ $+ b\|101\rangle_{012}\|0\rangle_3\|0\rangle_4$	$a\|001\rangle_{012}\|0\rangle_3\|0\rangle_4$ $+ b\|110\rangle_{012}\|0\rangle_3\|0\rangle_4$
$\|\psi\rangle_{s3} = \hat{U}_{CN(0,3)}\|\psi\rangle_{s2}$	$a\|000\rangle_{012}\|0\rangle_3\|0\rangle_4$ $+ b\|111\rangle_{012}\|1\rangle_3\|0\rangle_4$	$a\|100\rangle_{012}\|1\rangle_3\|0\rangle_4$ $+ b\|011\rangle_{012}\|0\rangle_3\|0\rangle_4$	$a\|010\rangle_{012}\|0\rangle_3\|0\rangle_4$ $+ b\|101\rangle_{012}\|1\rangle_3\|0\rangle_4$	$a\|001\rangle_{012}\|0\rangle_3\|0\rangle_4$ $+ b\|110\rangle_{012}\|1\rangle_3\|0\rangle_4$
$\|\psi\rangle_{s4} = \hat{U}_{CN(1,3)}\|\psi\rangle_{s3}$	$a\|000\rangle_{012}\|0\rangle_3\|0\rangle_4$ $+ b\|111\rangle_{012}\|0\rangle_3\|0\rangle_4$	$a\|100\rangle_{012}\|1\rangle_3\|0\rangle_4$ $+ b\|011\rangle_{012}\|1\rangle_3\|0\rangle_4$	$a\|010\rangle_{012}\|1\rangle_3\|0\rangle_4$ $+ b\|101\rangle_{012}\|1\rangle_3\|0\rangle_4$	$a\|001\rangle_{012}\|0\rangle_3\|0\rangle_4$ $+ b\|110\rangle_{012}\|0\rangle_3\|0\rangle_4$
$\|\psi\rangle_{s5} = \hat{U}_{CN(0,4)}\|\psi\rangle_{s4}$	$a\|000\rangle_{012}\|0\rangle_3\|0\rangle_4$ $+ b\|111\rangle_{012}\|0\rangle_3\|1\rangle_4$	$a\|100\rangle_{012}\|1\rangle_3\|1\rangle_4$ $+ b\|011\rangle_{012}\|1\rangle_3\|0\rangle_4$	$a\|010\rangle_{012}\|1\rangle_3\|0\rangle_4$ $+ b\|101\rangle_{012}\|1\rangle_3\|1\rangle_4$	$a\|001\rangle_{012}\|0\rangle_3\|0\rangle_4$ $+ b\|110\rangle_{012}\|0\rangle_3\|1\rangle_4$
$\|\psi\rangle_{s} = \hat{U}_{CN(0,4)}\|\psi\rangle_{s5}$	$a\|000\rangle_{012}\|0\rangle_3\|0\rangle_4$ $+ b\|111\rangle_{012}\|0\rangle_3\|0\rangle_4$	$a\|100\rangle_{012}\|1\rangle_3\|1\rangle_4$ $+ b\|011\rangle_{012}\|1\rangle_3\|1\rangle_4$	$a\|010\rangle_{012}\|1\rangle_3\|0\rangle_4$ $+ b\|101\rangle_{012}\|1\rangle_3\|0\rangle_4$	$a\|001\rangle_{012}\|0\rangle_3\|1\rangle_4$ $+ b\|110\rangle_{012}\|0\rangle_3\|1\rangle_4$
Corrective action	Nothing	$\|00\rangle_{34} \to \|11\rangle_{34}$; Apply the X gate to $q[0]$ to flip the 1st bit.	$\|00\rangle_{34} \to \|10\rangle_{34}$; Apply the X gate to $q[1]$ to flip the 2nd bit.	$\|00\rangle_{34} \to \|01\rangle_{34}$; Apply the X gate to $q[2]$ to flip the 3rd bit.

4.7.2 Phase flip error [15]

The phase flip error can be expressed $a|0\rangle + b|1\rangle \to a|1\rangle - b|0\rangle$. If we define $|+\rangle = \frac{1}{\sqrt{2}}(|0\rangle + |1\rangle)$ and

$|-\rangle = \frac{1}{\sqrt{2}}(|0\rangle - |1\rangle)$, then the phase flip will be $|+\rangle \to |-\rangle$ or $|-\rangle \to |+\rangle$.

This is essentially the same as the bit flip, and we can follow the parity check circuit used for the bit flip error correction. For the phase flip error, we need to construct $a|+++\rangle + b|---\rangle$. This can be performed by adding the H-gate to each qbit line in figure 4.18.

The form before the phase error detection is given by table 4.7

For a detailed argument of the bit flip and phase flip corrections, refer to [16].

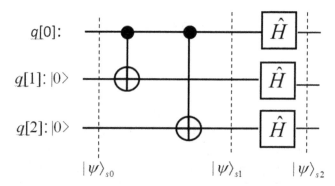

Figure 4.18. Adding redundancy for phase flip

Table 4.7. Step-by-step analysis of phase error detection.

$|\psi\rangle_{s0}$	$|\psi\rangle_{s1} = \hat{U}_{CN(0,2)}\hat{U}_{CN(0,1)}|\psi\rangle_{s0}$	$|\psi\rangle_{s2} = \hat{H}_0\hat{H}_1\hat{H}_2|\psi\rangle_{s1}$
$(a|0\rangle_0 + b|1\rangle_0)|0\rangle_1|0\rangle_2$	$a|0\rangle_0|0\rangle_1|0\rangle_2 + b|1\rangle_0|1\rangle_1|0\rangle_2$	$a|+\rangle_0|+\rangle_1|+\rangle_2 + b|-\rangle_0|-\rangle_1|-\rangle_2$

```
>>> from blueqat import Circuit
>>> import math
>>> #INVQFT|0000>=|0>
Circuit().swap[0,3].swap[1,2].h[0].cphase(-math.pi/2)[0,1].h[1].cphase(-math.pi/4)[0,2].cphase(-
math.pi/2)[1,2].h[2].cphase(-math.pi/8)[0,3].cphase(-math.pi/4)[1,3].cphase(-math.pi/2)[2,3].h[3].run()
array([0.25+0.j , 0.25+0.j , 0.25+0.j , 0.25+0.j , 0.25+0.j , 0.25+0.j,
    0.25+0.j, 0.25+0.j, 0.25+0.j , 0.25+0.j , 0.25+0.j , 0.25+0.j,
    0.25+0.j , 0.25+0.j , 0.25+0.j , 0.25+0.j])
>>> #INVQFT|0100>=|4>
Circuit().x[2].swap[0,3].swap[1,2].h[0].cphase(-math.pi/2)[0,1].h[1].cphase(-math.pi/4)[0,2].cphase(-
math.pi/2)[1,2].h[2].cphase(-math.pi/8)[0,3].cphase(-math.pi/4)[1,3].cphase(-math.pi/2)[2,3].h[3].run()
array([ 2.5000000e-01+0.j ,  1.5308085e-17-0.25j ,
    -2.5000000e-01+0.j ,  -1.5308085e-17+0.25j ,
     2.5000000e-01+0.j ,  1.5308085e-17-0.25j ,
    -2.5000000e-01+0.j ,  -1.5308085e-17+0.25j ,
     2.5000000e-01+0.j ,  1.5308085e-17-0.25j ,
    -2.5000000e-01+0.j ,  -1.5308085e-17+0.25j ,
     2.5000000e-01+0.j ,  1.5308085e-17-0.25j ,
    -2.5000000e-01+0.j ,  -1.5308085e-17+0.25j ])
>>> #INVQFT|1000>=|8>
Circuit().x[3].swap[0,3].swap[1,2].h[0].cphase(-math.pi/2)[0,1].h[1].cphase(-math.pi/4)[0,2].cphase(-
math.pi/2)[1,2].h[2].cphase(-math.pi/8)[0,3].cphase(-math.pi/4)[1,3].cphase(-math.pi/2)[2,3].h[3].run()
array([ 0.25+0.j, -0.25+0.j,  0.25+0.j, -0.25+0.j,  0.25+0.j, -0.25+0.j,
    0.25+0.j, -0.25+0.j,  0.25+0.j, -0.25+0.j,  0.25+0.j, -0.25+0.j,
    0.25+0.j, -0.25+0.j,  0.25+0.j, -0.25+0.j])
>>> #INVQFT|1100>=|12>
Circuit().x[2].x[3].swap[0,3].swap[1,2].h[0].cphase(-math.pi/2)[0,1].h[1].cphase(-math.pi/4)[0,2].cphase(-
math.pi/2)[1,2].h[2].cphase(-math.pi/8)[0,3].cphase(-math.pi/4)[1,3].cphase(-math.pi/2)[2,3].h[3].run()
array([ 2.5000000e-01+0.j ,  -1.5308085e-17+0.25j ,
    -2.5000000e-01+0.j ,  1.5308085e-17-0.25j ,
     2.5000000e-01+0.j ,  -1.5308085e-17+0.25j ,
    -2.5000000e-01+0.j ,  1.5308085e-17-0.25j ,
     2.5000000e-01+0.j ,  -1.5308085e-17+0.25j ,
    -2.5000000e-01+0.j ,  1.5308085e-17-0.25j ,
     2.5000000e-01+0.j ,  -1.5308085e-17+0.25j ,
    -2.5000000e-01+0.j ,  1.5308085e-17-0.25j ])
>>>
```

Figure 4.19. QFT^{-1} used for $M=15$ factoring.

```
>>> from blueqat import Circuit
>>> import math
>>> # INVQFT of |0001>=|1>
Circuit().x[0].swap[0,3].swap[1,2].h[0].cphase(-math.pi/2)[0,1].h[1].cphase(-math.pi/4)[0,2].cphase(-
math.pi/2)[1,2].h[2].cphase(-math.pi/8)[0,3].cphase(-math.pi/4)[1,3].cphase(-math.pi/2)[2,3].h[3].run()
array([ 2.50000000e-01+0.j,       2.30969883e-01-0.09567086j,
    1.76776695e-01-0.1767767j,   9.56708581e-02-0.23096988j,
    1.53080850e-17-0.25j,       -9.56708581e-02-0.23096988j,
   -1.76776695e-01-0.1767767j,  -2.30969883e-01-0.09567086j,
   -2.50000000e-01+0.j ,       -2.30969883e-01+0.09567086j,
   -1.76776695e-01+0.1767767j,  -9.56708581e-02+0.23096988j,
   -1.53080850e-17+0.25j,        9.56708581e-02+0.23096988j,
    1.76776695e-01+0.1767767j,   2.30969883e-01+0.09567086j ])
>>> # INVQFT of |0101>=|5>
Circuit().x[2].x[0].swap[0,3].swap[1,2].h[0].cphase(-math.pi/2)[0,1].h[1].cphase(-math.pi/4)[0,2].cphase(-
math.pi/2)[1,2].h[2].cphase(-math.pi/8)[0,3].cphase(-math.pi/4)[1,3].cphase(-math.pi/2)[2,3].h[3].run()
array([ 2.50000000e-01+0.j,      -9.56708581e-02-0.23096988j,
   -1.76776695e-01+0.1767767j,   2.30969883e-01+0.09567086j,
    1.53080850e-17-0.25j,       -2.30969883e-01+0.09567086j,
    1.76776695e-01+0.1767767j,   9.56708581e-02-0.23096988j,
   -2.50000000e-01+0.j ,        9.56708581e-02+0.23096988j,
    1.76776695e-01-0.1767767j,  -2.30969883e-01-0.09567086j,
   -1.53080850e-17+0.25j,        2.30969883e-01-0.09567086j,
   -1.76776695e-01-0.1767767j,  -9.56708581e-02+0.23096988j ])
>>> # INVQFT of |1001>=|9>
Circuit().x[3].x[0].swap[0,3].swap[1,2].h[0].cphase(-math.pi/2)[0,1].h[1].cphase(-math.pi/4)[0,2].cphase(-
math.pi/4)[1,2].h[2].cphase(-math.pi/8)[0,3].cphase(-math.pi/4)[1,3].cphase(-math.pi/2)[2,3].h[3].run()
array([ 2.50000000e-01+0.j,      -2.30969883e-01+0.09567086j,
    1.76776695e-01-0.1767767j,  -9.56708581e-02+0.23096988j,
    1.53080850e-17-0.25j,        9.56708581e-02+0.23096988j,
   -1.76776695e-01-0.1767767j,   2.30969883e-01+0.09567086j,
   -2.50000000e-01+0.j ,        2.30969883e-01-0.09567086j,
   -1.76776695e-01+0.1767767j,   9.56708581e-02-0.23096988j,
   -1.53080850e-17+0.25j,       -9.56708581e-02-0.23096988j,
    1.76776695e-01+0.1767767j, - 2.30969883e-01-0.09567086j ])
>>> # INVQFT of |1101>=|13>
Circuit().x[3].x[2].x[0].swap[0,3].swap[1,2].h[0].cphase(-math.pi/2)[0,1].h[1].cphase(-math.pi/4)[0,2].cphase(-
math.pi/2)[1,2].h[2].cphase(-math.pi/8)[0,3].cphase(-math.pi/4)[1,3].cphase(-math.pi/2)[2,3].h[3].run()
array([ 2.50000000e-01+0.j,       9.56708581e-02+0.23096988j,
   -1.76776695e-01+0.1767767j,  -2.30969883e-01-0.09567086j,
    1.53080850e-17-0.25j,        2.30969883e-01-0.09567086j,
    1.76776695e-01+0.1767767j,  -9.56708581e-02+0.23096988j,
   -2.50000000e-01+0.j ,       -9.56708581e-02-0.23096988j,
    1.76776695e-01-0.1767767j,   2.30969883e-01+0.09567086j,
   -1.53080850e-17+0.25j,       -2.30969883e-01+0.09567086j,
   -1.76776695e-01-0.1767767j,   9.56708581e-02-0.23096988j ])
>>>
```

Figure 4.19. (Continued.)

```
>>> from blueqat import Circuit
>>> import math
>>> # INVQFT of |0010>=|2>
Circuit().x[1].swap[0,3].swap[1,2].h[0].cphase(-math.pi/2)[0,1].h[1].cphase(-math.pi/4)[0,2].cphase(-
math.pi/2)[1,2].h[2].cphase(-math.pi/8)[0,3].cphase(-math.pi/4)[1,3].cphase(-math.pi/2)[2,3].h[3].run()
array([ 2.50000000e-01+0.j  ,        1.76776695e-01-0.1767767j ,
        1.53080850e-17-0.25j  ,      -1.76776695e-01-0.1767767j ,
       -2.50000000e-01+0.j  ,        -1.76776695e-01+0.1767767j ,
       -1.53080850e-17+0.25j ,        1.76776695e-01+0.1767767j ,
        2.50000000e-01+0.j  ,         1.76776695e-01-0.1767767j ,
        1.53080850e-17-0.25j ,       -1.76776695e-01-0.1767767j ,
       -2.50000000e-01+0.j   ,       -1.76776695e-01+0.1767767j ,
       -1.53080850e-17+0.25j ,        1.76776695e-01+0.1767767j ])
>>> # INVQFT of |0110>=|6>
Circuit().x[2].x[1].swap[0,3].swap[1,2].h[0].cphase(-math.pi/2)[0,1].h[1].cphase(-math.pi/4)[0,2].cphase(-
math.pi/2)[1,2].h[2].cphase(-math.pi/8)[0,3].cphase(-math.pi/4)[1,3].cphase(-math.pi/2)[2,3].h[3].run()
array([ 2.50000000e-01+0.j  ,       -1.76776695e-01-0.1767767j ,
       -1.53080850e-17+0.25j ,        1.76776695e-01-0.1767767j ,
       -2.50000000e-01+0.j  ,         1.76776695e-01+0.1767767j ,
        1.53080850e-17-0.25j ,       -1.76776695e-01+0.1767767j ,
        2.50000000e-01+0.j  ,        -1.76776695e-01-0.1767767j ,
       -1.53080850e-17+0.25j ,        1.76776695e-01-0.1767767j ,
       -2.50000000e-01+0.j  ,         1.76776695e-01+0.1767767j ,
        1.53080850e-17-0.25j ,       -1.76776695e-01+0.1767767j ])
>>> # INVQFT of |1010>=|10>
Circuit().x[3].x[1].swap[0,3].swap[1,2].h[0].cphase(-math.pi/2)[0,1].h[1].cphase(-math.pi/4)[0,2].cphase(-
math.pi/2)[1,2].h[2].cphase(-math.pi/8)[0,3].cphase(-math.pi/4)[1,3].cphase(-math.pi/2)[2,3].h[3].run()
array([ 2.50000000e-01+0.j  ,       -1.76776695e-01+0.1767767j ,
        1.53080850e-17-0.25j ,        1.76776695e-01+0.1767767j ,
       -2.50000000e-01+0.j  ,         1.76776695e-01-0.1767767j ,
       -1.53080850e-17+0.25j ,       -1.76776695e-01-0.1767767 j ,
        2.50000000e-01+0.j  ,        -1.76776695e-01+0.1767767j ,
        1.53080850e-17-0.25j ,        1.76776695e-01+0.1767767j ,
       -2.50000000e-01+0.j  ,         1.76776695e-01-0.1767767j ,
       -1.53080850e-17+0.25j ,        1.76776695e-01-0.1767767j ])
>>> # INVQFT of |1110>=|14>
Circuit().x[3].x[2].x[1].swap[0,3].swap[1,2].h[0].cphase(-math.pi/2)[0,1].h[1].cphase(-math.pi/4)[0,2].cphase(-
math.pi/2)[1,2].h[2].cphase(-math.pi/8)[0,3].cphase(-math.pi/4)[1,3].cphase(-math.pi/2)[2,3].h[3].run()
array([ 2.50000000e-01+0.j  ,        1.76776695e-01+0.1767767j ,
       -1.53080850e-17+0.25j ,       -1.76776695e-01+0.1767767j ,
       -2.50000000e-01+0.j  ,        -1.76776695e-01-0.1767767j ,
        1.53080850e-17-0.25j ,         1.76776695e-01-0.1767767j ,
        2.50000000e-01+0.j   ,        1.76776695e-01+0.1767767j ,
       -1.53080850e-17+0.25j  ,      -1.76776695e-01+0.1767767j ,
       -2.50000000e-01+0.j  ,        -1.76776695e-01-0.1767767j ,
        1.53080850e-17-0.25j ,         1.76776695e-01-0.1767767j ])
>>>
```

Figure 4.19. (Continued.)

```
>>> from blueqat import Circuit
>>> import math
>>> #INVQFT|0011>=|3>
Circuit().x[0].x[1].swap[0,3].swap[1,2].h[0].cphase(-math.pi/2)[0,1].h[1].cphase(-math.pi/4)[0,2].cphase(-
math.pi/2)[1,2].h[2].cphase(-math.pi/8)[0,3].cphase(-math.pi/4)[1,3].cphase(-math.pi/2)[2,3].h[3].run()
array([ 2.50000000e-01+0.j ,      9.56708581e-02-0.23096988j ,
       -1.76776695e-01-0.1767767j , -2.30969883e-01+0.09567086j ,
       -1.53080850e-17+0.25j ,     2.30969883e-01+0.09567086j ,
        1.76776695e-01-0.1767767j , -9.56708581e-02-0.23096988j ,
       -2.50000000e-01+0.j ,      -9.56708581e-02+0.23096988j ,
        1.76776695e-01+0.1767767j , 2.30969883e-01-0.09567086j ,
        1.53080850e-17+0.25j ,     -2.30969883e-01-0.09567086j ,
       -1.76776695e-01+0.1767767j , 9.56708581e-02+0.23096988j ])
>>> #INVQFT|0111>=|7>
Circuit().x[0].x[1].x[2].swap[0,3].swap[1,2].h[0].cphase(-math.pi/2)[0,1].h[1].cphase(-math.pi/4)[0,2].cphase(-
math.pi/2)[1,2].h[2].cphase(-math.pi/8)[0,3].cphase(-math.pi/4)[1,3].cphase(-math.pi/2)[2,3].h[3].run()
array([ 2.50000000e-01+0.j ,      -2.30969883e-01-0.09567086j ,
        1.76776695e-01+0.1767767j , -9.56708581e-02-0.23096988j ,
       -1.53080850e-17+0.25j ,     9.56708581e-02-0.23096988j ,
       -1.76776695e-01+0.1767767j , 2.30969883e-01-0.09567086j ,
       -2.50000000e-01+0.j ,       2.30969883e-01+0.09567086j ,
       -1.76776695e-01-0.1767767j , 9.56708581e-02+0.23096988j ,
        1.53080850e-17+0.25j ,     -9.56708581e-02+0.23096988j ,
        1.76776695e-01-0.1767767j , -2.30969883e-01+0.09567086j ])
>>> #INVQFT|1011>=|11>
Circuit().x[0].x[1].x[2].swap[0,3].swap[1,2].h[0].cphase(-math.pi/2)[0,1].h[1].cphase(-math.pi/4)[0,2].cphase(-
math.pi/2)[1,2].h[2].cphase(-math.pi/8)[0,3].cphase(-math.pi/4)[1,3].cphase(-math.pi/2)[2,3].h[3].run()
array([ 2.50000000e-01+0.j ,      -2.30969883e-01-0.09567086j ,
        1.76776695e-01+0.1767767j , -9.56708581e-02-0.23096988j ,
       -1.53080850e-17+0.25j ,     9.56708581e-02-0.23096988j ,
       -1.76776695e-01+0.1767767j , 2.30969883e-01-0.09567086j ,
       -2.50000000e-01+0.j ,       2.30969883e-01+0.09567086j ,
       -1.76776695e-01-0.1767767j , 9.56708581e-02+0.23096988j ,
        1.53080850e-17-0.25j ,     -9.56708581e-02+0.23096988j ,
        1.76776695e-01-0.1767767j , -2.30969883e-01+0.09567086j ])
>>> #INVQFT|1111>=|15>
Circuit().x[0].x[1].x[2].x[3].swap[0,3].swap[1,2].h[0].cphase(-math.pi/2)[0,1].h[1].cphase(-math.pi/4)[0,2].cphase(-
math.pi/2)[1,2].h[2].cphase(-math.pi/8)[0,3].cphase(-math.pi/4)[1,3].cphase(-math.pi/2)[2,3].h[3].run()
array([ 2.50000000e-01+0.j ,      2.30969883e-01+0.09567086j ,
        1.76776695e-01+0.1767767j , 9.56708581e-02+0.23096988j ,
       -1.53080850e-17+0.25j ,     -9.56708581e-02+0.23096988j ,
       -1.76776695e-01+0.1767767j , -2.30969883e-01+0.09567086j ,
       -2.50000000e-01+0.j ,       -2.30969883e-01-0.09567086j ,
       -1.76776695e-01-0.1767767j , -9.56708581e-02-0.23096988j ,
        1.53080850e-17-0.25j ,     9.56708581e-02-0.23096988j ,
        1.76776695e-01-0.1767767j , 2.30969883e-01-0.09567086j ])
>>>
```

Figure 4.19. (Continued.)

References

[1] Bernhardt C 2019 *Quantum Computing for Everyone* (Cambridge, MA: The MIT Press)
[2] Mermin N D *007* Quantum Computer Science An Introduction (Cambridge: Cambridge University Press)
[3] Nielsen M A and Chuang I L 2010 *Quantum Computation and Quantum Information* (Cambridge: Cambridge University Press)

[4] Candela D 2015 Undergraduate computational physics projects on quantum computing *Am. J. Phys.* **83** 688

[5] Gharibian S 2015 *Lecture 6: Deutsch's Algorithm* Virginia Commonwealth University) (www.people.vcu.edu/~sgharibian/courses/CMSC491..)

[6] *blueqat-tutorials/100_deutsch.ipynb at master · Blueqat/blueqat-tutorials · GitHub*

[7] Qiskit 2018 *Deutsch–Jozsa Algorithm* (*Deutsch–Jozsa Algorithm (qiskit.org)*)

[8] Ye A 2020 *Grover's Algorithm—Quantum Computing* https://medium.com/swlh/grovers-algorithm-quantum-computing-1171e826bcfb

[9] Berkley U C 2014 C/CS/Phys C191 Quantum Fourier Transform 11/05–11/07

[10] Cho S 2018 *Fourier Transform and its Applications Using Microsoft EXCEL® IOP Concise Physics* (San Rafael, CA: Morgan & Claypool)

[11] IBM 2021 *Shor's Algorithm* (IBM Quantum)

[12] Shor P 2015 *What is Shor's factoring algorithm?—YouTube* Physics World. Also watch P. Shor 2021 *The Story of Shor's Algorithm, Straight From the Source | Peter Shor—YouTube*

[13] Salton G, Simon D and Lin C 2021 *Exploring Simon's Algorithm with Daniel Simon* AWS Quantum Computing Blog

[14] Zhou B 2020 Quantum Error Correction Codes (University of Arizona) https://wp.optics.arizona.edu/opti646/wp-content/…

[15] Mykhailova M 2020 *Phase Flip Error Correction on State |0>* (Quantum Computing Stack Exchange)

[16] Calderbank A R and Shor P W 1995 Good quantum error-correcting codes exist *Phys. Rev. A* **54** 1098

IOP Publishing

Quantum Computation and Quantum Information
Simulation using Python
A gentle introduction
Shinil Cho

Chapter 5

Quantum information: entanglement and teleportation

While transmitting a qbit, it is easily collapsed due to external factors such as 'thermal noise.' Bennett *et al* proposed the idea of transmitting a qbit without collapsing the quantum state [1]. Their idea uses the spooky behavior of a pair of entangled spins (section 1.5.1), and is called quantum teleportation. In this transmission scheme, the transmitter and the receiver share a pair of entangled quantum states, and then the receiver reconstructs the qbit based on the information it received about what the transmitter finds from the qbit entangled with the receiver. Analyzing a simple quantum teleportation circuit should help us understand their idea on how the entanglement is applied in this circuit.

The quantum entanglement is such a peculiar concept that even very experienced physicists were very skeptical. How do we know the quantum entanglement is true? John Bell described how to determine numerically whether quantum entanglement is real. He proposed an inequality that should be satisfied if the concept of the quantum entanglement is true rather than the classical interpretation of a pair of two spin states [2]. It involves observations of entangled spin pairs in the coordinate systems different from the z-direction. There are many articles that prove Bell's inequality with very rigorous mathematics. However, in this book, observing spin states from a direction other than the z-direction is described with the projection operators (section 3.2).

We also discuss the superdense (or ultra-high density) coding scheme in this chapter. This process is a quantum communication protocol often considered as the reverse operation of quantum teleportation. In this scheme, if a sender wants to send a classical 2-bit information (00, 01, 10, or 11), all the sender needs to do is to send a single qbit to a receiver who shares an entangled resource with the sender. This protocol was proposed by Bennett and Wiesner [3].

doi:10.1088/978-0-7503-3963-6ch5

5.1 Bell's inequality

John Bell discussed the probabilities of observing spin states in different observation coordinate frames, which showed the different values of probability between the classical interpretation and the quantum interpretation of the quantum entanglement. Suppose Alice and Bob have an entangled spin state,

$$|\psi\rangle = \frac{1}{\sqrt{2}}(|0_A\rangle|0_B\rangle + |1_A\rangle|1_B\rangle) \tag{5.1}$$

in the default 'vertical' measurement coordinate frame where the orthonormal basis of the spin state is given by $\{|0>, |1>\}_{\theta=0}$. Bell came up with the idea of measuring the probability of observing the entanglement in three different measurement coordinate frames whose orthonormal bases are given by $\{|0'>, |1'>\}_{\theta=0}$, $\{|0'>, |1'>\}_{\theta=2\pi/3}$, and $\{|0'>, |1'>\}_{\theta=4\pi/3}$, respectively. The classical interpretation of the spin pairs and the consequence of the different quantum entanglement outcome probabilities, which can be measured experimentally.

5.1.1 Classical interpretation of entangled states

There are eight possible spin states, $2^3=8$, that Alice and Bob observe when each of them selects one measurement frame from the three different measurement frames of $\theta = 0$, $2\pi/3$, and $4\pi/3$. Recall that, in the classical interpretation, the spin states are determined before any observation. Table 5.1 below shows the possible spin states in the classical interpretation where Alice and Bob observe when they select one of the three measurement frames where their measurement frames at an angle θ are denoted as A_θ and B_θ. Recall that the spin states are pre-determined in the classical interpretation, and we can write down all possible spin states, and then find the probability of a particular spin state. From the result of table 5.1, the probability that Alice and Bob observe the same spin directions, $|0_A>|0_B>$ or $|1_A>|1_B>$, turns out to be at least 5/9 with the classical interpretation.

5.1.2 Quantum entanglement

First, we need to depict the entangled states in the measurement coordinate frames at $\theta=2\pi/3$ and $\theta=4\pi/3$. In order to obtain them, we need to find how the entanglement will be changed if the measurement coordinate frames changes from the orthonormal basis to another orthonormal basis, say $\{|b_0>, |b_1>\}$. This is equivalent to rotation of the vertical coordinate frame by given angles. We will show that, for the entangled state given by equation (5.1),

$$|\psi\rangle = \frac{1}{\sqrt{2}}(|0_A\rangle|0_B\rangle + |1_A\rangle|1_B\rangle) = \frac{1}{\sqrt{2}}(|b_{A0}\rangle|b_{B0}\rangle + |b_{A1}\rangle|b_{B1}\rangle) \tag{5.2}$$

where the suffixes A and B stand for Alice and Bob.

Table 5.1. Classical interpretation of spin entanglement.

{Possible pre-determined spin states} Probability Alice and Bob observe the same spin states.	A_0	B_θ	$A_{2\pi/3}$	B_θ	$A_{4\pi/3}$	B_θ
{|0>, |0>, |0>} **Probability =1**	|0>	**B_0:** |0> **$B_{2\pi/3}$:** |0>' **$B_{4\pi/3}$:** |0>'	|0>'	**B_0:** |0> **$B_{2\pi/3}$:** |0>' **$B_{4\pi/3}$:** |0>'	|0>'	**B_0:** |0> **$B_{2\pi/3}$:** |0>' **$B_{4\pi/3}$:** |0>'
{|0>, |0>, |1>} **Probability = 5/9**	|0>	**B_0:** |0> **$B_{2\pi/3}$:** |0>' $B_{4\pi/3}$: |1>'	|0>'	**B_0:** |0> **$B_{2\pi/3}$:** |0>' $B_{4\pi/3}$: |1>'	|1>'	|0>: |0> **$B_{2\pi/3}$:** |0>' **$B_{4\pi/3}$:** |1>'
{|0>, |1>, |0>} **Probability = 5/9**	|0>	**B_0:** |0> $B_{2\pi/3}$: |1>' **$B_{4\pi/3}$:** |0>'	|1>'	B_0: |0> **$B_{2\pi/3}$:** |1>' **$B_{4\pi/3}$:** |0>'	|0>'	**B_0:** |0> $B_{2\pi/3}$: |1>' **$B_{4\pi/3}$:** |0>'
{|1>, |0>, |0>} **Probability = 5/9**	|1>	**B_0:** |1> $B_{2\pi/3}$: |0>' $B_{4\pi/3}$: |0>'	|0>'	B_0: |1> **$B_{2\pi/3}$:** |0>' **$B_{4\pi/3}$:** |0>'	|0>'	|0>: |1> **$B_{2\pi/3}$:** |0>' **$B_{4\pi/3}$:** |0>'
{|0>, |1>, |1>} **Probability = 5/9**	|0>	**B_0:** |0> $B_{2\pi/3}$: |1>' **$B_{4\pi/3}$:** |1>'	|1>'	B_0: |0> **$B_{2\pi/3}$:** |1>' **$B_{4\pi/3}$:** |1>'	|1>'	|0>: |0> **$B_{2\pi/3}$:** |1>' **$B_{4\pi/3}$:** |1>'
{|1>, |0>, |1>} **Probability = 5/9**	|1>	**B_0:** |1> $B_{2\pi/3}$: |0>' **$B_{4\pi/3}$:** |1>'	|0>'	**B_0:** |1> $B_{2\pi/3}$: |0>' **$B_{4\pi/3}$:** |1>'	|1>'	**B_0:** |1> $B_{2\pi/3}$: |0>' **$B_{4\pi/3}$:** |1>'
{|1>, |1>, |0>} **Probability = 5/9**	|1>	**B_0:** |1> **$B_{2\pi/3}$:** |1>' $B_{4\pi/3}$: |0>'	|1>'	**B_0:** |1> **$B_{2\pi/3}$:** |1>' $B_{4\pi/3}$: |0>'	|0>'	**B_0:** |1> **$B_{2\pi/3}$:** |1>' $B_{4\pi/3}$: |0>'
{|1>, |1>, |1>} **Probability = 1**	|1>	**B_0:** |1> **$B_{2\pi/3}$:** |1>' **$B_{4\pi/3}$:** |1>'	|1>'	**B_0:** |1> **$B_{2\pi/3}$:** |1>' **$B_{4\pi/3}$:** |1>'	|1>'	**B_0:** |1> **$B_{2\pi/3}$:** |1>' **$B_{4\pi/3}$:** |1>'

Proof:

Let $\left\{|b_0\rangle = \begin{bmatrix} a \\ b \end{bmatrix}, |b_1\rangle = \begin{bmatrix} c \\ d \end{bmatrix}\right\}$, then we can express $\left\{|0\rangle = \begin{bmatrix} 1 \\ 0 \end{bmatrix}, |1\rangle = \begin{bmatrix} 0 \\ 1 \end{bmatrix}\right\}$ using the new orthonormal basis:

$$|0\rangle = \hat{P}_{|b_0\rangle}|0\rangle + \hat{P}_{|b_1\rangle}|0\rangle = a|b_0\rangle + c|b_1\rangle, \tag{5.3}$$

and

$$|1\rangle = \hat{P}_{|b_0\rangle}|0\rangle + \hat{P}_{|b_1\rangle}|1\rangle = b|b_0\rangle + d|b_1\rangle$$

where we used the projection operators (equations (1.7) and (1.19)). Because

$$|0_A\rangle|0_B\rangle = (a|b_{A0}\rangle + c|b_{A1}\rangle)|0_B\rangle = a|b_{A0}\rangle|0_B\rangle + c|b_{A1}\rangle|0_B\rangle = |b_{A0}\rangle a|0_B\rangle + |b_{A1}\rangle c|0_B\rangle,$$

and

$$|1_A\rangle|1_B\rangle = (b|b_{A0}\rangle + d|b_{A1}\rangle)|1_B\rangle = b|b_{A0}\rangle|1_B\rangle + d|b_{A1}\rangle|1_B\rangle = |b_{A0}\rangle b|1_B\rangle + |b_{A1}\rangle d|1_B\rangle,$$

the entangled state becomes

$$|\psi\rangle = \frac{1}{\sqrt{2}}(|0_A\rangle|0_B\rangle + |1\rangle_A\rangle|1_B\rangle) = \frac{1}{\sqrt{2}}(|b_{A0}\rangle a|0_B\rangle + |b_{A1}\rangle c|0_B\rangle + |b_{A0}\rangle b|1_B\rangle + |b_{A1}\rangle d|1_B\rangle)$$

$$= \frac{1}{\sqrt{2}}[|b_{A0}\rangle(a|0\rangle + b|1\rangle)_B + |b_{A1}\rangle(c|0\rangle + d|1\rangle)_B] = \frac{1}{\sqrt{2}}[|b_{A0}\rangle|b_{B0}\rangle + |b_{A1}\rangle|b_{B1}\rangle]. \tag{5.4}$$

In order to find the measurement probabilities when, for example Alice uses $\{|0>, |1>\}_{\theta=0}$ and Bob uses $\{|0>', |1>'\}_{\theta=2\pi/3}$, we rotate Bob's measurement coordinate frames. Recall that the measurement coordinate frame rotation is given by equation (1.13), we obtain

$$\text{If } \theta = 2\pi/3, \ |0\rangle' = \begin{bmatrix} \cos(\pi/3) \\ -\sin(\pi/3) \end{bmatrix} = \begin{bmatrix} 1/2 \\ -\sqrt{3}/2 \end{bmatrix} \text{ and } |1\rangle' = \begin{bmatrix} \sin(\pi/3) \\ \cos(\pi/3) \end{bmatrix} = \begin{bmatrix} \sqrt{3}/2 \\ 1/2 \end{bmatrix}; \tag{5.5}$$

and

$$\text{if } \theta = 4\pi/3, \ |0\rangle'' = \begin{bmatrix} \cos(2\pi/3) \\ -\sin(2\pi/3) \end{bmatrix} = \begin{bmatrix} -1/2 \\ -\sqrt{3}/2 \end{bmatrix} \text{ and } |1\rangle'' = \begin{bmatrix} \sin(2\pi/3) \\ \cos(2\pi/3) \end{bmatrix} = \begin{bmatrix} \sqrt{3}/2 \\ -1/2 \end{bmatrix}. \tag{5.6}$$

There are four possible selections of the measurement frames. We can calculate the probability of observing the same spin states from each case.

(1) Both Alice and Bob use the $\theta=0$ frames. If both observe the same spin states, $|0_A >|0_B>$ or $|1_A>|1_B>$,

$$|\psi\rangle = \frac{1}{\sqrt{2}}(|0_A\rangle|0_B\rangle + |1_A\rangle|1_B\rangle). \tag{5.7}$$

Thus, the probability that both Alice and Bob observe the same spin orientation is 1/2 each.

(2) Alice uses the $\theta=0$ frame and measures her spin first, then Bob uses the $\theta=2\pi/3$ frame to measure his spin. In order to find the measurement

probabilities, we need to express $|0_B\rangle$ and $|1_B\rangle$ using the orthonormal basis of the $\theta=2\pi/3$ frame:

$$|0_B\rangle = \hat{P}_{|0'\rangle}|0_B\rangle + \hat{P}_{|1'\rangle}|0_B\rangle = \frac{1}{2}|0'_B\rangle + \frac{\sqrt{3}}{2}|1'_B\rangle, \qquad (5.8)$$

and

$$|1_B\rangle = \hat{P}_{|0'\rangle}|1_B\rangle + \hat{P}_{|1'\rangle}|1_B\rangle = -\frac{\sqrt{3}}{2}|0'_B\rangle + \frac{1}{2}|1'_B\rangle. \qquad (5.9)$$

Thus, we obtain

$$|\psi\rangle = \frac{1}{\sqrt{2}}(|0_A\rangle|0_B\rangle + |1_A\rangle|1_B\rangle)$$

$$= \frac{1}{\sqrt{2}}\left[|0_A\rangle\left(\frac{1}{2}|0'\rangle + \frac{\sqrt{3}}{2}|1'\rangle\right)_B + |1_A\rangle\left(-\frac{\sqrt{3}}{2}|0'\rangle + \frac{1}{2}|1'\rangle\right)_B\right] \qquad (5.10)$$

$$= \frac{1}{\sqrt{2}}\left[\frac{1}{2}|0_A\rangle|0'_B\rangle + \frac{\sqrt{3}}{2}|0_A\rangle|1'_B\rangle - \frac{\sqrt{3}}{2}|1_A\rangle|0'_A\rangle + \frac{1}{2}|1_A\rangle|1'_B\rangle\right].$$

Therefore, the probability that both Alice and Bob observe the same spin orientation, $|0_A\rangle|0_B'\rangle$ or $|1_A\rangle|1_B'\rangle$, is 1/8 each.

(3) Alice uses the $\theta=0$ frame and measures her spin first, then Bob uses the $\theta=4\pi/3$ frame to measure his spin. In order to find the measurement probabilities, we need to express $|0_B\rangle$ and $|1_B\rangle$ using the orthonormal basis of the $\theta=4\pi/3$ frame:

$$|0_B\rangle = \hat{P}_{|0''\rangle}|0_B\rangle + \hat{P}_{|1''\rangle}|0_B\rangle = -\frac{1}{2}|0''_B\rangle + \frac{\sqrt{3}}{2}|1''_B\rangle, \qquad (5.11)$$

and

$$|1_B\rangle = \hat{P}_{|0''\rangle}|1_B\rangle + \hat{P}_{|1''\rangle}|1_B\rangle = -\frac{\sqrt{3}}{2}|0''_B\rangle - \frac{1}{2}|1''_B\rangle. \qquad (5.12)$$

Thus,

$$|\psi\rangle = \frac{1}{\sqrt{2}}(|0_A\rangle|0_B\rangle + |1_A\rangle|1_B\rangle)$$

$$= \frac{1}{\sqrt{2}}\left[|0_A\rangle\left(-\frac{1}{2}|0''_B\rangle + \frac{\sqrt{3}}{2}|1''_B\rangle\right) + |1_A\rangle\left(-\frac{\sqrt{3}}{2}|0''_B\rangle - \frac{1}{2}|1''_B\rangle\right)\right] \qquad (5.13)$$

$$= \frac{1}{\sqrt{2}}\left[-\frac{1}{2}|0\rangle_A|0''_B\rangle| + \frac{\sqrt{3}}{2}|0\rangle_A|1''_B\rangle - \frac{\sqrt{3}}{2}|1\rangle_A|0''_A\rangle - \frac{1}{2}|1\rangle_A|1''_B\rangle\right].$$

Therefore, the probability that both Alice and Bob observe the same spin orientation, $|0_A\rangle|0_B\rangle$ or $|1_A\rangle|1_B\rangle$, is also 1/8 each.

(4) If Alice uses the $\theta=2\pi/3$ frame and observes her spin first and Bob uses the $\theta=4\pi/3$ frame to observe his spin, the quantum state jumps to $|0'_A\rangle|0'_B\rangle + |1'_A\rangle|1'_B\rangle$ after Alice's measurement.

$$|0'_B\rangle = \hat{P}_{|0''\rangle}|0'_B\rangle + \hat{P}_{|1''\rangle}|1'_B\rangle$$

$$= \left[-\frac{1}{2} \quad -\frac{\sqrt{3}}{2} \right] \begin{bmatrix} \frac{1}{2} \\ -\frac{\sqrt{3}}{2} \end{bmatrix} |0''_B\rangle + \left[\frac{\sqrt{3}}{2} \quad -\frac{1}{2} \right] \begin{bmatrix} \frac{1}{2} \\ -\frac{\sqrt{3}}{2} \end{bmatrix} |1''_B\rangle \tag{5.14}$$

$$= \frac{1}{2}|0''_B\rangle + \frac{\sqrt{3}}{2}|1''_B\rangle.$$

Thus, we obtain

$$|\psi\rangle = \frac{1}{\sqrt{2}}(|0'_A\rangle|0'_B\rangle + |1'_A\rangle|1'_B\rangle)$$

$$= \frac{1}{\sqrt{2}}\left[|0'_A\rangle\left(\frac{1}{2}|0''\rangle + \frac{\sqrt{3}}{2}|1''\rangle\right)_B + |1'_A\rangle\left(-\frac{\sqrt{3}}{2}|0''\rangle + \frac{1}{2}|1''\rangle\right)_B \right] \tag{5.15}$$

$$= \frac{1}{\sqrt{2}}\left[\frac{1}{2}|0'_A\rangle|0''_B\rangle + \frac{\sqrt{3}}{2}|0'_A\rangle|1''_B\rangle - \frac{\sqrt{3}}{2}|1'_A\rangle|0''_A\rangle + \frac{1}{2}|1'_A\rangle|1''_B\rangle \right].$$

The probability that both Alice and Bob observe the same spin orientation, $|0_A\rangle|0_B\rangle$ or $|1_A\rangle|1_B\rangle$, is 1/8 each again. Notice that this qbit state is essentially the same as that of the case (2). Thus, the number of different configurations of measurement frames are 3.

From the above result, we can conclude that, in 1/3 of the measurement events, if Alice observes $|0\rangle$, then Bob always observes $|0\rangle$, i.e., the probability is equal to 1/2 when Alice uses the same measurement frames. In 2/3 of the measurement events, they observe $|0\rangle|0\rangle$ with the probability of 1/8 when they observe the spins in the different frames. In other words, the total probability of observing the entangled state $|0\rangle|0\rangle$ is (1/3) •1 + (2/3)•(1/8) = 0.25. The same is true for observing $|1\rangle|1\rangle$. Thus, the probabilities of observing the same spin directions, $|0\rangle|0\rangle$ or $|1\rangle|1\rangle$, is 0.5, which is smaller than the 5/9 of the classical consequence! In 1972, the experiment (Bell test) was conducted by Freedman and Clauser and supported the quantum entanglement [4].

Figure 5.1 is the demonstration of Bell's inequality using Blueqat. In these circuits, Bell's gate described in chapter 3 creates entangled states. Then the entangled states will be observed using two different coordinate frames. In all cases, the total probabilities of observing the entangled states, $|0\rangle|0\rangle$ or $|1\rangle|1\rangle$, is indeed 0.5!

Blueqat codes and outputs

Blueqat codes and outputs

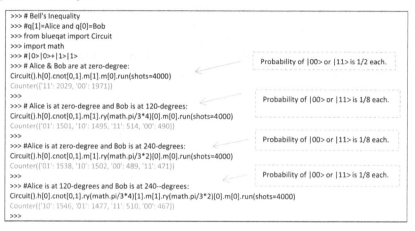

```
>>> # Bell's Inequality
>>> #q[1]=Alice and q[0]=Bob
>>> from blueqat import Circuit
>>> import math
>>> #|0>|0>+|1>|1>
>>> # Alice & Bob are at zero-degree:
Circuit().h[0].cnot[0,1].m[1].m[0].run(shots=4000)
Counter({'11': 2029, '00': 1971})
>>>
>>> # Alice is at zero-degree and Bob is at 120-degrees:
Circuit().h[0].cnot[0,1].m[1].ry(math.pi/3*4)[0].m[0].run(shots=4000)
Counter({'01': 1501, '10': 1495, '11': 514, '00': 490})
>>>
>>> #Alice is at zero-degree and Bob is at 240-degrees:
Circuit().h[0].cnot[0,1].m[1].ry(math.pi/3*2)[0].m[0].run(shots=4000)
Counter({'01': 1538, '10': 1502, '00': 489, '11': 471})
>>>
>>> #Alice is at 120-degrees and Bob is at 240--degrees:
Circuit().h[0].cnot[0,1].ry(math.pi/3*4)[1].m[1].ry(math.pi/3*2)[0].m[0].run(shots=4000)
Counter({'10': 1546, '01': 1477, '11': 510, '00': 467})
>>>
```

Probability of |00> or |11> is 1/2 each.

Probability of |00> or |11> is 1/8 each.

Probability of |00> or |11> is 1/8 each.

Probability of |00> or |11> is 1/8 each.

Figure 5.1. Bells' inequality.

5.2 Quantum teleportation

We follow Bennett's idea. Suppose Alice wants to send a qbit ($|\psi>=\alpha|0>+\beta|1>$) to Bob. Notice that Alice does not know what this qbit is because once she observes it, it will be collapsed to either $|0>$ or $|1>$, and we hence cannot clone a qbit (section 1.4). Nevertheless, Alice can 'transmit' it to Bob using an EPR pair!

The flow of operation is as follows:

(1) Alice has two electrons of $|\psi>=\alpha|0>+\beta|1>$ and $|0>$ whereas Bob has one electron of $|0>$;

(2) Alice and Bob create an entangled state of their $|0>$s using Bell's gate (section 3.3.5);

(3) Alice operates the reverse Bell gate and then observes the two electrons to acquire one of the four states of $|00>$, $|01>$, $|10>$, and $|11>$; and

(4) depending on what Alice observes, Bob's quantum bit jumps to the corresponding state due to the entanglement.

Figure 5.2 shows this scheme using the Bell gate and the reverse Bell gate.

Step-by-step analysis

$|\psi\rangle_{s1} = |\psi\rangle|0\rangle|0\rangle$ where $|\psi\rangle = \alpha|0\rangle + \beta|1\rangle$,　　This is a superposed state.

$$|\psi\rangle_{s2} = \frac{1}{\sqrt{2}}|\psi\rangle(|00\rangle + |11\rangle) = \frac{1}{\sqrt{2}}(\alpha|0\rangle + \beta|1\rangle)(|0\rangle + |1\rangle)|0\rangle$$

$$= \frac{1}{\sqrt{2}}(\alpha|0\rangle + \beta|1\rangle)(|00\rangle + |10\rangle),$$

$$|\psi\rangle_{s3} = \hat{U}_{CN}|\psi\rangle_{s2} = \frac{1}{\sqrt{2}}(\alpha|0\rangle + \beta|1\rangle)\hat{U}_{CN}(|00\rangle + |10\rangle)$$　　|0> and |0> are entangled.

$$= \frac{1}{\sqrt{2}}(\alpha|0\rangle + \beta|1\rangle)(|00\rangle + |11\rangle) = \frac{1}{\sqrt{2}}\left[\alpha|0\rangle(|00\rangle + |11\rangle) + \beta|1\rangle(|00\rangle + |11\rangle)\right],$$

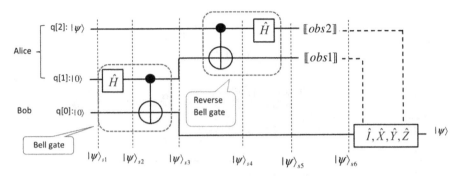

Figure 5.2. Quantum teleportation.

Table 5.2. Quantum teleportation.

[obs2]	[obs1]	Bob's state before gate	Gate	Bob's state after gate
0	0	$\alpha|0\rangle + \beta|1\rangle$	$\hat{I}$	$\alpha|0\rangle + \beta|1\rangle = |\psi\rangle$
0	1	$\alpha|1\rangle + \beta|0\rangle$	$\hat{X}$	
1	0	$\alpha|0\rangle - \beta|1\rangle$	$\hat{Z}$	
1	1	$\alpha|1\rangle - \beta|0\rangle$	$\hat{Y}$	

$$|\psi\rangle_{s4} = \hat{U}_{CN}|\psi\rangle_{s3} = \frac{1}{\sqrt{2}}\hat{U}_{CN}(\alpha|0\rangle + \beta|1\rangle)(|00\rangle + |11\rangle)$$

$$= \frac{1}{\sqrt{2}}[\alpha|0\rangle(|00\rangle + |11\rangle) + \beta|1\rangle(|10\rangle + |01\rangle)],$$

$$|\psi\rangle_{s5} = \hat{H}|\psi\rangle_{s4} = \frac{1}{\sqrt{2}}\hat{H}[\alpha|0\rangle(|00\rangle + |11\rangle) + \beta|1\rangle(|10\rangle + |01\rangle)]$$

$$= \frac{1}{2}[\alpha(|0\rangle + |1\rangle)(|00\rangle + |11\rangle) + \beta(|0\rangle - |1\rangle)(|10\rangle + |01\rangle)],$$

$$|\psi\rangle_{s6} = |\psi\rangle_{s5} = \frac{1}{2}[\alpha(|0\rangle + |1\rangle)(|00\rangle + |11\rangle) + \beta(|0\rangle - |1\rangle)(|10\rangle + |01\rangle)]$$

$$= \frac{1}{2}[|00\rangle(\alpha|0\rangle + \beta|1\rangle) + |01\rangle(\alpha|1\rangle + \beta|0\rangle) + |10\rangle(\alpha|0\rangle - \beta|1\rangle) + |11\rangle(\alpha|1\rangle - \beta|0\rangle)].$$

Therefore, the possibility of observations [obs2] for $q[2]$ and [obs1] for $q[1]$ should be one of the following. Alice informs Bob of her observations so that Bob can apply the proper gate to produce |ψ> as shown in table 5.2

Blueqat codes and outputs
The following codes listed in figure 5.3 output the typical results where $q[2]$ is the code to be teleported. $q[1]$ from Alice and $q[0]$ from Bob is entangled. When $q[2]=|0>$ is

```
>>> from blueqat import Circuit
>>>#Quantum teleportation
>>> #q[2]=|0>, q[1]=|0>, q[0]=|0>
Circuit().h[1].cx[1,0].cx[2,1].h[2].cx[1,0].cz[2,0].m[:].run(shots=100)
Counter({'000': 28, '001': 27, '010': 25, '011': 20})
>>> #q[2]=|1>, q[1]=|0>, q[0]=|0>
Circuit().x[2].h[1].cx[1,0].cx[2,1].h[2].cx[1,0].cz[2,0].m[:].run(shots=100)
Counter({'111': 35, '101': 23, '100': 21, '110': 21})
>>> #q[2]=(|0>+|1>)/SQRT(2), q[1]=|0>, q[1]=|0>
Circuit().h[2].h[1].cx[1,0].cx[2,1].h[2].cx[1,0].cz[2,0].m[:].run(shots=100)
Counter({'100': 16, '111': 15, '110': 15, '011': 13, '001': 12, '000': 11, '010': 10, '101': 8})
>>>
```

Figure 5.3. Quantum teleportation.

transported, regardless of the measurement status of $q[1]$ and $q[0]$, the output $q[2]$ continues to be $|0\rangle$. Similarly when $q[2]=|1\rangle$ is transported, the output $q[2]$ continues to be $|1\rangle$. When $q[2]=|0\rangle+|1\rangle$ is transported, the output $q[2]$ is $|0\rangle$ or $|1\rangle$ with equal probabilities.

Quantum teleportation has been successfully demonstrated in several countries with much more sophisticated setups. For implementation of experimental quantum teleportation, refer to [5].

5.3 Superdense coding

The superdense coding takes advantage of qbits [6]. The superdense coding circuit using Bell gates (section 3.3.6) and reverse Bell gates (section 3.3.7) is shown in figure 5.4 where Alice wants to send one of classical 2-bite data (00, 01, 10, or 11).

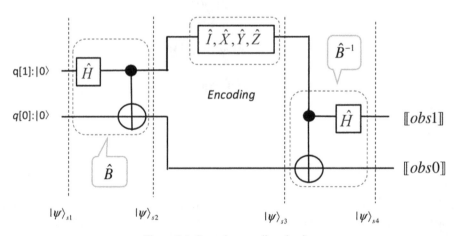

Figure 5.4. Superdense coding circuit.

Step-by-step analysis

$$|\psi\rangle_{s1} = |0\rangle|0\rangle,$$

$$|\psi\rangle_{s2} = \hat{B}|\psi\rangle_{s1} = \frac{1}{\sqrt{2}}(|00\rangle + |11\rangle).$$

Blueqat codes and outputs

Table 5.3 lists the superdense coding scheme. From this table, the superdense coding is straightforward and self-explanatory as shown in figure 5.5.

Table 5.3. Superdense coding scheme.

Classical two bits to be sent	Encoding gate	$|\psi\rangle_{s3}$	$|\psi\rangle_{s4} = \hat{B}^{-1}|\psi\rangle_{s3}$	$[obs1\&0]$
00	$\hat{I}$	$\hat{I}|\psi\rangle_{s2}$ $= \frac{1}{\sqrt{2}}(|00\rangle + |11\rangle)$	$\hat{B}^{-1}|\psi\rangle_{s3}$ $= \hat{B}^{-1}\left[\frac{1}{\sqrt{2}}(|00\rangle + |11\rangle)\right] = |00\rangle$	00
01	$\hat{Z}$	$\hat{Z}|\psi\rangle_{s2}$ $= \frac{1}{\sqrt{2}}(|00\rangle - |11\rangle)$	$\hat{B}^{-1}|\psi\rangle_{s3}$ $= \hat{B}^{-1}\left[\frac{1}{\sqrt{2}}(|00\rangle - |11\rangle)\right] = |01\rangle$	01
10	$\hat{X}$	$\hat{X}|\psi\rangle_{s2}$ $= \frac{1}{\sqrt{2}}(|10\rangle + |01\rangle)$	$\hat{B}^{-1}|\psi\rangle_{s3}$ $= \hat{B}^{-1}\left[\frac{1}{\sqrt{2}}(|10\rangle + |01\rangle)\right] = |10\rangle$	10
11	$\hat{Y}$	$\hat{Y}|\psi\rangle_{s2}$ $= \frac{i}{\sqrt{2}}(|00\rangle - |01\rangle)$	$\hat{B}^{-1}|\psi\rangle_{s3}$ $= \hat{B}^{-1}\left[\frac{i}{\sqrt{2}}(|10\rangle - |01\rangle)\right] = i|11\rangle$	11

```
>>> from blueqat import Circuit
>>> #Superdense Coding
>>> #Classical bits to be sent = 00
Circuit().h[1].cx[1,0].cx[1,0].h[1].m[:].run(shots=100)
Counter({'00': 100})
>>> #Classical bits to be sent = 01
Circuit().h[1].cx[1,0].z[1].cx[1,0].h[1].m[:].run(shots=100)
Counter({'01': 100})
>>> #Classical bits to be sent = 10
Circuit().h[1].cx[1,0].x[1].cx[1,0].h[1].m[:].run(shots=100)
Counter({'10': 100})
>>> #Classical bits to be sent = 11
Circuit().h[1].cx[1,0].y[1].cx[1,0].h[1].m[:].run(shots=100)
Counter({'11': 100})
>>>
```

Figure 5.5. Superdense coding.

References

[1] Bennett C 1998 Quantum information theory *IEEE Trans. Inform. Theory* **44** (https://academia.edu/3729794)
[2] Bell J S 1964 On the Einstein–Podolsky–Rosen Paradox *Phys. Physiq. Физика.* **1** 195–200
[3] Bennet C and Wiesner S 1992 Communication via one- and two-particle operators on Einstein–Podolsky–Rosen states *Phys Rev Lett.* **69** 2881–84
[4] Freedman S J and Clauser J F 1972 Experimental test of local hidden-variable theories *Phys. Rev. Lett.* **28** 938–41
[5] Jeewandara T 2022 *Experimental quantum teleportation of propagating microwaves* (phys.org)
[6] Jadhav N 2021 *Understanding superdense coding. An in-depth explanation and tutorial of… | by Nidhi Jadhav | Geek Culture | Medium*

Chapter 6

Quantum cryptography
(quantum key distribution)

In this chapter, we explore a quantum scheme of transmitting a secret cryptography key, BB84, proposed by Charles Bennett and Gilles Brassard [1, 2]. It is the first protocol using the quantum property of photons that allows two people to exchange a secret key for cryptography. One of the major problems with secret key cryptography is the logistical issue of how to transmit the key from one party to the other without allowing access to an attacker. The essential presumption of the scheme of a secret key-based cryptography is that such a key must be known only by the sender and the receiver. If a third party obtains the key, the whole scheme is useless. Thus, it is critical that the secret key scheme cannot be tampered with. The BB84 protocol does not prevent such tampering but detects it so that only untampered code can be transmitted and used.

Readers may wonder how a secret key encodes and decodes a message. Not many articles and books demonstrate the scheme although it is interesting to know. Thus, this book explains how a secret key works.

6.1 Cryptography using a secret key

Suppose a plain text, $\{M\}=\{M_1, M_2, ..., M_n\}$ where M_i are ASCII codes, needs to be understandable by only the sender and the receiver. The text will be encrypted using a secret key. First we convert the plain text into corresponding binary codes, which are then encrypted with a predetermined binary number sequence, $\{K\}=\{K_1, K_2, ..., K_n\}$, as an encryption key. Such a key can be generated using random numbers. The easiest way to encrypt the binary text is a classical XOR-gate. Readers should note that the protocol below is similar to the secret key we describe in Simon's algorithm (section 4.6). The encrypted code, $\{C\}=\{C_1, C_2, ..., C_n\}$, is defined as $C_i= \text{XOR}(K_i, M_i)$ by a sender where $i=1, 2, ..., n$. The receiver decodes

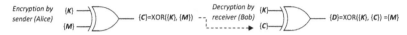

Figure 6.1. Cryptography using a secret key.

{*C*} using the same secret key by applying *D*i=XOR(*K*i, *C*i) =*M*i as shown in figure 6.1.

Example: Two characters 'Q' and 'M' will be encrypted as an example, Their binary code of 'QM' is {*M*}={01010001 01001101}. Suppose the secret key is created {*K*}={10011100 10100111} by a random sequence of 1 and 0, then the encrypted code of {*M*} is {*C*}={11001101 11101010}, which is now 'iê'. In order to perform decryption by a receiver, Bob, the XOR-gate can be applied again. The output {*D*} is {01010001 01001101}. This is the correctly decrypted characters 'QM'. In this way, the original text {*M*} can be encrypted and decrypted with the same key.

In the above scheme, the key {*K*} is common to both the sender and receiver but must be secret to anyone else. If someone else knows the key, the encrypted code can be decoded. A secret key in this encryption and decryption scheme must not be tampered with by a third person. If the sender and the receiver are unaware of tampering by a third person, the key is no longer secret. Therefore, there must be some mechanism in the transmission protocol that can detect if there is a third party.

The BB84 protocol is a method of transmitting a secret key through a quantum channel based on the non-cloning principle of quantum states (section 1.4). If a third party, say Eve, attempts to tamper with the channel to obtain the key, Alice and Bob can detect Eve's activity because the key will be altered by tampering.

6.2 Photon-based qbit

As the original BB84 protocol used a photon-based qbit (section 1.6), we also use photon states to transmit a secret key. BB84 used photons of four different polarization directions, 0°, 45°, 90°, and 135°. Here, we use the horizontally polarized photon ($|\leftrightarrow\rangle$) as |0> and the vertically polarized photon ($|\updownarrow\rangle$) as |1>. For the other photons polarized along 45° and 135°, we denote then as $|\nearrow\rangle$ and $|\searrow\rangle$, respectively.

We use two optical devices to create and observe the photon-based qbit: polarization rotators and polarization beam splitters. Polarization beam splitters are devices that have two separate outputs for a horizontally polarized photon ($|\leftrightarrow\rangle$) and for a vertically polarized photon ($|\updownarrow\rangle$). Notice that, for an arbitrary polarized photon, $|\psi\rangle = a|\leftrightarrow\rangle + b|\updownarrow\rangle$, which represents a photon-based qbit, the beam splitter outputs both $|\leftrightarrow\rangle$ with the probability $|a|^2$ and $|\updownarrow\rangle$ with the probability $|b|^2$. In particular, for +45° ($|\nearrow\rangle$) and 135° ($|\searrow\rangle$) polarizations, the probability to output $|\leftrightarrow\rangle$ or $|\updownarrow\rangle$ is equal and 1/2. Figure 6.2 shows a functional diagram of a polarization beam splitter.

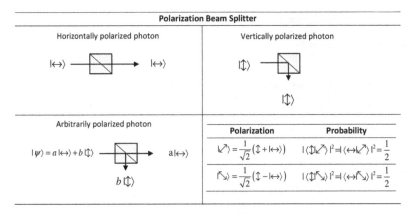

Figure 6.2. Polarization beam splitter.

6.3 BB84 protocol

The BB84 protocol is now described using the secret key $\{K\}=\{10011100\ 10100111\}$ used above. Suppose a sender, Alice, wants to send the secret key to a receiver, Bob, through a photon quantum channel. The quantum channel for the transmission can be established by using photon polarizations. Alice converts the array of 16 binary numbers to a set of 16 photons where the $\leftrightarrow$ is assigned to the binary number 0 and $\updownarrow$ is the binary number 1. There are four possible directions of polarization: $0°$ ($\leftrightarrow$), $45°$ ($\nearrow$), $90°$ ($\updownarrow$), and $135°$ (or $-45°$) ($\nwarrow$). Notice if $\nearrow$ or $\nwarrow$ are to be observed ambiguously, polarization rotators can be used to rotate them to $|\updownarrow\rangle$ or $|\leftrightarrow\rangle$, respectively and then use the splitter.

In this book, we assume that Alice uses a rotator ($+45°$) and a combination of the rotator and a beam splitter whereas Bob uses another type of rotator ($-45°$) and a combination of the rotator and a beam splitter.

Transmission of secret key

(1) **Sender (Alice):** Alice converts each bit of the secret key $\{K\}$ into a single photon state in the following steps. She prepares a set of randomly configured the polarization bean splitters (+) and sets of a polarization rotator ($+45°$) plus a splitter ($\times$). Following the transmission rule of the table 6.1, Alice determines the polarization states of 16 photons and produces the photons, and sends the photon to Bob.

Suppose she gets the following polarization device array for the code key $\{K\}$. Now, each binary number of the code key $\{K\}$ is assigned to the horizontal or the vertical polarization, i.e., $\leftrightarrow$ is assigned as 0 and $\updownarrow$ is assigned as 1.

Bit #	1	2	3	4	5	6	7	8	9	10	11	12	13	14	15	16
Binary numbers	1	0	0	1	1	1	0	0	1	0	1	0	0	1	1	1
Photon states	$\updownarrow$	$\leftrightarrow$	$\leftrightarrow$	$\updownarrow$	$\updownarrow$	$\updownarrow$	$\leftrightarrow$	$\leftrightarrow$	$\updownarrow$	$\leftrightarrow$	$\updownarrow$	$\leftrightarrow$	$\leftrightarrow$	$\updownarrow$	$\updownarrow$	$\updownarrow$
Alice's device array	+	+	×	+	×	×	+	+	+	×	×	+	+	×	+	×

Table 6.1. Polarization rules for transmission.

Direction of linear polarization	↔	↕	↗	↘
Polarization beam splitter only (+)	↔	↕	↗	↘
Polarization rotator (+45°) plus splitter (×)	↗	↘	↕	↔

Table 6.2. Polarization rule for reception.

Direction of linear polarization	↔	↕	↘	↘
Polarization beam splitter only (+)	↔	↕	↗	↘
Polarization rotator (−45°) plus splitter (×')	↘	↘	↔	↕

In the above table, the first bit (1) of the code key assigns ↕ and produces a photon ↕ through the polarization splitter only (+), and the fifth bit (1) assigns ↕ and produces a photon of ↗ through the polarization rotator plus the splitter (×), and so forth. The produced photon set is now ready to send to Bob.

Bit #	1	2	3	4	5	6	7	8	9	10	11	12	13	14	15	16
Photons sent	↕	↔	↗	↕	↘	↘	↔	↔	↕	↗	↘	↔	↔	↘	↕	↘

(2) **Receiver (Bob):** Bob prepares his set of randomly configured polarization bean splitters (+) and sets of a polarization rotator (−45°) plus a splitter (×') as the photon receiver array. The receiving rule is given in table 6.2.

(3) Suppose he builds the following array.

Bit #	1	2	3	4	5	6	7	8	9	10	11	12	13	14	15	16
Bob's device array	+	×'	+	+	+	×'	×'	×'	+	+	×'	+	+	+	×'	+

Bob conducts quantum measurement using his detector array and determines the secret key, using the receiving rule shown in table 6.1.

Bit #	1	2	3	4	5	6	7	8	9	10	11	12	13	14	15	16
Photons received	↕	↔	↗	↕	↘	↘	↔	↔	↕	↗	↘	↔	↔	↘	↕	↘
Bob's device array	+	×'	+	+	+	×'	×'	×'	+	+	×'	+	+	+	×'	+
Observed binary numbers	1	1or0	1or0	1	1or0	1	1or0	1or0	1	1or0	1	0	0	1or0	1or0	1or0

Bob received the configuration of Alice's device array via a classical communication method such as the telephone. Then, he compares it with his array configuration to confirm that the 1st, 4th, 6th, 9th, 11th, 12th, and 13th are the same. Alice and Bob can share the binary number in the matched array number, which is observed unambiguously, as the secret key. It is important to note that Bob only shares with Alice what detector array numbers are the same so that no one else knows the actual device arrays of Alice and Bob. By repeating this procedure, Alice and Bob can construct a secret key of 16-bit or more.

(4) Now, suppose Eve taps the communication using Eve's own detector array shown below. Eve taps Alice's photon set, and produce another set of photon states.

Bit #	1	2	3	4	5	6	7	8	9	10	11	12	13	14	15	16
Photons received	↗	↔	↗	↕	↗	↗	↘	↘	↕	↘	↘	↔	↔	↘	↕	↗
Eve's device array	×'	+	+	×'	×'	×'	+	+	×'	+	+	×'	+	+	+	×'
Acquired photon states	↗	↘	↗	↗	↕	↗	↘	↔	↗	↘	↗	↘	↔	↗	↘	↕

Because of the non-cloning principle, Eve can only send the measured photon set to Bob. Bob conducts a quantum measurement with what he receives.

Bit #	1	2	3	4	5	6	7	8	9	10	11	12	13	14	15	16
Photons received	↗	↔	↗	↗	↕	↕	↔	↔	↗	↗	↘	↘	↔	↘	↕	↕
Bob's device array	+	×'	+	+	+	×'	×'	×'	+	+	×'	+	+	+	×'	+
Acquired photon states	↗	↘	↗	↗	↕	↗	↘	↘	↗	↗	↕	↘	↔	↘	↗	↕
Observed binary numbers	1or0	1or0	1or0	**1or0**	0	**1or** 0	1or0	1or 0	**1or0**	1or0	**1**	1 or 0	**0**	1or0	1or0	1

Bob receives information on Alice's polarizer array via a classical communication method such as a telephone. Bob compares the array information with his to confirm the 1st, 4th, 6th, 9th, 11th, 12th, and 13th are the same. Here, the 11th and 13th are accidentally matched while the other bits (1st, 4th, 6th, 9th, and 12th) are not. Thus, Bob detects Eve's presence.

What is the probability that Bob and Alice are not aware of Eve's tapping assuming all arrays are constructed in a random fashion? The probability that Eve selects one of the polarization detectors (+ or ×) is 1/2 per bit. The probability that the Bob and Alice do not know Eve is given by

[The probability Eve selects the same detector as Bob does]
+ [the probability Eve selects a wrong detector and Bob selects a wrong detector]
= (1/2) + (1/2)(1/2) = 3/4 per bit.

Therefore, if there are n-bit code, the probability that Alice and Bob are unaware of Eve is $(3/4)^n$, and the probability that Alice and Bob are aware of Eve is given by $1-(3/4)^n = 0.99999999$ if $n=64$!

References

[1] Bennett C, Giles B and Ekert A 1992 *Quantum Cryptography* (Scientific American) October, pp 50–7
[2] Gisin N, Ribordy G, Tittle R and Zbinde H 2002 *Quantum Cryptography* (American Physical Society) (https://cdn.journals.aps.org/files/RevModPhys.74.145.pdf)

Appendix A

Commercial quantum computers

A.1 Implementation of qbits

How can we physically create qbits? There are several methods for producing practical qbits [1–3]. Here, we briefly describe how qbits can be created.

(1) Qbits can be provided by nuclear spins within dissolved molecules in a solution and probes with the nuclear magnetic resonance (NMR). Pulse-NMR can flip a nuclear spin by an arbitrary angle to implement quantum gates.

(2) A two-dimensional electron system where an electron trapped on the surface of liquid helium can be a qbit.

(3) A Josephson junction, which is a superconducting wire loop, can behave like a qbit by assigning a value to the direction that the current flows around an electrical circuit made by the superconducting wire loop.

(4) Ions can be captured in a magnetic field just like a quadrupole mass spectrometer. The trapped ions are used as qbits whose states are assigned based on their energy states between the ground state and an excited state.

(5) The silicon spin qubits are made of semiconductor material and are used to contain and manipulate electrons. Each qubit is composed of a single electron trapped in a tiny chamber called a double quantum dot. By applying a microwave to the electron, the electron spin can be flipped up or down to assign the qubit a quantum state of 1 or 0.

(6) Single particles of light can be prepared in a quantum superposition of two different colors. For example, a Ca-atom emits two orthogonal light waves (purple (423 nm) and green (551 nm)) in opposite directions when the Ca-atom is excited in a particular energy level by a laser. This system can represent qbits with optical filters and detectors.

A.2 Commercial quantum computers

In order to establish practical quantum computation, we need many qbits to communicate with each other, although today's quantum computers contain tens of qbits made using the technologies described above. Nevertheless, we now have commercial quantum computers available. We summarize the current four major companies that offers qbit-based computer services in alphabetical order. Refer to their websites for more information.

    (1) Amazon Braket [4]: Amazon offers a quantum computing service with different types of quantum computers that users can select. Its SDK is called Braket and the platform is AWS.

    (2) Google Quantum AI [5]: Google also offers a cloud-based quantum computing service. Its SDK is a Python library called Cirq. The cloud-based platform is Google Cloud Platform (GCP).

    (3) IBM Q system [6]: this is an industry first initiative to build universal quantum computers for business, engineering and science. Its software development ki (SDK) is Qiskill.

    (4) Microsoft [7]: this is an open cloud quantum computing system. Its SDK is open-source, called Q# quantum programing language, and the platform is called Azure.

    (5) There are many emerging hardware companies in this field, including Ion Q [8] and D-Wave [9]. Ion Q used a specialized chip called a linear ion trap (section 1.6). D-wave uses quantum annealing processors for optimization problems. Quantum annealing finds low-energy states of a problem utilizing quantum-tunneling.

We are witnessing rapid progress in the development of quantum computers. The latest development of a silicon based qbit achieved a major step in increasing the scalability of quantum computers [10]. We may soon have a desktop quantum computer for just $5000 [11]!

References

[1] Shea S B 2020 Creating the Heart of a Quantum Computer: Developing Qubits (Creating the Heart of a Quantum Computer: Developing Qubits | Department of Energy) https://www.energy.gov/science/articles/creating-heart-quantum-computer-developing-qubits

[2] Tarinm A,, Cieslak A and Bucley I 2015 *How to Make a Qubit* https://cs.usfca.edu/~jcchubb/StudentWork/Qubits.pdf

[3] DiVincenzo D P 2002 *The Physical Implementation of Quantum, Computation* https://arxiv.org/abs/quant-ph/0002077

[4] Quantum Computing Service—Amazon Braket–Amazon Web Services https://aws.amazon.com/braket/

[5] Google Quantum AI https://quantumai.google/

[6] What is Quantum Computing? | IBM https://www.ibm.com/topics/quantum-computing

[7] Microsoft Quantum Overview | Microsoft Azure https://azure.microsoft.com/en-us/solutions/quantum-computing/

[8] IonQ | Trapped Ion Quantum Computing https://ionq.com/

[9]   D-Wave Systems | The Practical Quantum Computing https://dwavesys.com/
[10] Riken Researchers Demonstrated a Triple-Qubit, Silicon-Based Quantum Computing Mechanism—Inside Quantum Technology https://www.insidequantumtechnology.com/news-archive/riken-researchers-demonstrated-a-triple-qubit-silicon-based-quantum-computing-mechanism/
[11] A Desktop Quantum Computer for Just $5,000 | Discover Magazine https://www.discovermagazine.com/technology/a-desktop-quantum-computer-for-just-usd5-000